PHILOSOPHICAL DARWINISM

PHILOSOPHICAL DARWINISM

On the origin of knowledge by means of natural selection

Peter Munz

London and New York

First published 1993
by Routledge
11 New Fetter Lane, London EC4P 4EE

Simultaneously published in the USA and Canada
by Routledge
a division of Routledge, Chapman and Hall, Inc.
29 West 35th Street, New York, NY 10001

Typeset in 10/12 pt Baskerville
Datakey
Printed in Great Britain by
T J Press (Padstow) Ltd, Padstow, Cornwall.

British Library Cataloguing in Publication Data

A catalogue record for this book is available from the British Library.

Library of Congress Cataloging in Publication Data

Munz, Peter—
Philosophical Darwinism : on the origin of knowledge by means of natural selection/Peter Munz.
p. cm.
Includes bibliographical references and index.
1. Knowledge, Theory of. 2. Evolution. 3. Natural selection.
I. Title.
BD177.M86 1993
121—dc20
92–19364

ISBN 0–415–08602–7

CONTENTS

PREFACE

Though the history of ideas has its own momentum, I am writing as a historian who has noticed that something has been overlooked, so that it seems necessary to apply a little corrective pressure to that momentum. It is over a hundred years ago that Darwin suggested that we revise our philosophy by discarding Locke and amending Plato. He was looking at the matter from the vantage point of biology, and this book is an attempt to work out the philosophical consequences of biology. Although Darwin's suggestions are well known and frequently quoted, philosophers have so far taken little notice of them, and biology has not yet replaced physics as the centre of attention. In the middle of the twentieth century, when it was realised that Bacon's New Atlantis had turned out to be Max Weber's Iron Cage, inhabited by Riesman's Lonely Crowd, and that the view that scientific theories have a partial observational interpretation by means of correspondence rules should never have become the Received View, philosophers started to move away from the long tradition of modernism, which had stretched from Bacon and Locke to the early Wittgenstein and to Carnap. Disillusioned with modernism, they turned a blind eye to the implications of biology and veered instead towards the post-modern relativism of Kuhn, the post-modern post-structuralism of Foucault, Derrida and Lyotard or the post-modern pragmatism of Rorty, and are showing unending and increasing interest in the obfuscations of Heidegger. It is perfectly true that modernism was too sanguine in its belief that language was a completely transparent and neutral medium which would allow a rigid and easy distinction between reality and fiction, science and literature. But the nowadays widely popular conclusions – that science is nothing but a form of literature, that the dogmatic beliefs of primitive cultures are as valid and explanatory as the highly general theories of cosmopolitan science, that all authors, including the living ones, are 'dead' and that readers are free to read anything they like into a text, that statements are parochial affirmations or chants which mirror nothing and that the problem of reference is obviated because nothing refers to anything – are simply irrational over-reactions to the glib pretensions of modernism. I will argue instead that the philosophical consequences of biology amount to a *rational* post-modern alternative to these unreasonable post-modern fashions.

In so far as this book is about the philosophy of science, I would like to state that it is my view that such a philosophy can neither prescribe anything to science nor answer questions which science cannot answer, but that it is interpretative in the sense that we need philosophy in order to find out what it is that science is telling us.

Since I am a historian and not a biologist, I came to the enormous explanatory potential of Darwinism indirectly, by two different but ultimately converging routes. As a historian in search of explanations of modern society, I found that my researches were driven back at first to the sixteenth century and then to the middle ages, and from there, to ancient history and prehistory and, finally, to the evolution of the earliest hominid societies, where I encountered Darwin, evolution and biology. On the more theoretical side, my attention was drawn to Darwin and evolution by Karl Popper's *Objective Knowledge* of 1972.

Parts of this book were written in Tuscania in my cousin Miriam Frank's house from which a minute fraction of the past from which we are descended is actually visible to the naked eye. In addition, this book owes a great debt to discussions with colleagues and friends: Erich Geiringer, my mentor in biology, John Roberts, Peter Webster, Bob Tristram and Harvey Franklin, all in Wellington; the late W. W. Bartley III (Stanford), Bernulf Kanitscheider and Gerhard Vollmer (Giessen), Gerard Radnitzky (Trier), Ian Jarvie and Jagdish Hattiangadi (York University, Toronto), Michael Ruse (Guelph University), Paul Levinson (New York), Paul Hoffmann (Tübingen), Anthony O'Hear (University of Bradford), Peter Wilson (University of Otago) and Donald T. Campbell (Lehigh University). It is only fair to add that some of them warned me that my conclusions were too bold. I wish to thank the Rev. Canon Derek Stanesby for inviting me to take part in Consultations at St George's House, Windsor Castle; Hans Albert for inviting me to the European Forum in Alpbach; Eugene Kamenka for invitations to the Unit of the History of Ideas at the Australian National University in Canberra; the Direzione Generale per le Relazioni Culturali del Ministero degli Affari Esteri of Italy for inviting me to a Conference in Venice; and the Victoria University of Wellington (New Zealand) for their continuing support of my researches. I specially thank Barbro Harris and her staff in the Reference Department and Alex Heatley in the Computer Centre; and Tim Naylor, who taught me how to manipulate my software. Last but not least, I owe a great debt to the Fritz Thyssen Stiftung of Köln in Germany for several generous grants which enabled me to work in German libraries.

Tuscania (Provincia di Viterbo),
Largo della Pace 7,
June 1992

ACKNOWLEDGEMENTS

Most of Chapter 2 first appeared in G. Radnitzky, ed., *Centripetal Forces in the Sciences*, New York, Paragon House, 1987, and one section of Chapter 4 in Fred D'Agostino and I. C. Jarvie, eds, *Freedom and Rationality*, Dordrecht, Kluwer Academic Publishers, 1989. I am grateful to the publishers concerned for permission to reprint this material.

INTRODUCTION: COGNITIVE CONDITIONS

Towards the end of the *Tractatus*, Wittgenstein remarked (6.44) 'not *how* the world is, is the mystical, but *that* it is'. One cannot be quite sure what he intended by the expression 'mystical', and given his later arguments in *Philosophical Investigations*, there can be no meaning attached to the question as to what he, as the author of the remark, 'intended' by it. One is, therefore, allowed a certain latitude in the reading of this text, and I will take it that he meant that the existence of the world is a *mysterium tremendum*; but that the *way* it is, is not. Nobody can deny that the mere existence of the world is a total mystery. But Wittgenstein was wrong in thinking that the *way* it is, is not.

On the face of it, Wittgenstein's distinction seems valid enough. We realise that the existence of the world is incomprehensible and that there can be no explanation as to why the Big Bang – if that was what it was – ever occurred. But how loud it was, that its echo can be detected at the present day, that the temperatures rose and then fell within the first three seconds, and so on – these are matters which appear to be explicable and, therefore, intelligible. But as soon as we probe, we discover that science, which informs us about how these things happened after the Big Bang, is, contrary to Wittgenstein's expectation, not self-explanatory. We cannot simply say: wait until science finds out. Science can only find out how the world works and *how* it is if one has an understanding of the concepts of truth, meaning and reference – to mention only the most crucial problems. Once this is seen, knowledge as to *how* the world is, is as problematical and, in many senses, as mysterious as *that* it is.

At the outset, I will take it that there is no absolute difference between what we understand as 'science' and knowledge in general. As Thomas Kuhn put it, 'science [is] our surest example of sound knowledge'.[1] I will therefore equate 'knowledge' with what we take to be 'specially refined knowledge', namely science; and refer to unrefined knowledge as 'false knowledge'. The matter is well put by Karl Popper:

> The whole trouble ... [is] due to the mistaken belief that *scientific knowledge* was an especially strict or august kind of knowledge ... [I

> explained] that in the usual sense of 'know', whenever I *know* that it is raining, it must be true that it is raining; for if it is not true, then I simply cannot know that it is raining, however sincerely I may believe that I know it. In this sense of the word, 'knowledge' always means 'true and certain knowledge'; and 'to know' means, in addition, to be in possession of *sufficient reason* for holding that our knowledge is true and certain ... There is no such thing as scientific knowledge in this sense. If, nonetheless, we chose to label the results of our scientific endeavours with the customary name 'scientific knowledge', then we ought to be clear that scientific knowledge ... [is] not a species of knowledge; least of all a species distinguished by a high degree of solidity and certainty. On the contrary, measured by the high standards of criticism, 'scientific knowledge' always remained sheer guesswork – although guesswork controlled by criticism and experiment.[2]

Criticism and experiment are public activities, and it follows, therefore, that while 'knowledge' can be private, secret and esoteric, knowledge as 'science' cannot. Science is, essentially, public knowledge, and an important difference between science and knowledge is not, as is widely believed, the certainty of the former, but the fact that the former is public while the latter need not be.[3]

There is certainly a mystery surrounding knowledge. We have legitimate doubts about what counts as meaning, reference and truth, and, given the vast disagreements as to whether these terms are to be taken in the sense of an operationalist conventionalism or in a more realistic sense – to mention only the two most divergent possibilities – it is a real mystery that we benefit so much from knowledge and that there is so much knowledge around.

Knowledge, then, is not self-explanatory. 'Whatever can be known, can be known by means of science', Quine wrote, echoing Wittgenstein. If the reference of all terms used in knowledge went without saying, the entire truth about the world, and, what is more, the *only* truth about the world would indeed be contained in the sentences we call knowledge. But such reference does not go without saying; and the mere fact that even if such clear reference could be established, it would have to be established by thinking other than 'scientific' thinking and would have to be meta-cognitive, proves that knowledge cannot depend on itself. For these reasons, the proper province of philosophy is the philosophy of knowledge. This should not be taken in a normative sense in which philosophers sit down and prescribe to scientists how they ought to proceed; but in an interpretative sense, in which philosophy seeks to find out what precisely it is we can learn from knowledge.

Philosophers in the later twentieth century, have, for the most part, misunderstood this situation. Having watched how one branch of knowledge after another has been removed from what used to be called 'natural philosophy' and having found that in the end even so philosophical a pursuit as psychology or the science of mind was removed from them, they became eager to define an area which could be seen as the subject matter of philosophy, of which they would continue to enjoy a monopoly, which they could call their own and which would ensure – and this was the great driving force – the continued existence of departments of philosophy in universities in which they could find jobs. In this way, they have made themselves unwittingly superfluous and have neglected that the proper study of mankind is man; or, in this case, that the proper study of philosophy is meta-cognition or the philosophy of science and knowledge.

There has indeed always been a lot of knowledge around. Ever since the earliest hominid days, men and women must have known what to eat and what not to eat, how to catch animals and what benefits one could derive from watching the movements of the sun. Some of that knowledge was true; some of it was false. But the important feature of such early knowledge was that, among other things, it was used for non-cognitive purposes. It was used as social cement, as a social bonding principle. Societies can be held together by shared beliefs, and such shared beliefs are expressed by the practice of ritual. To do its job as a social bonding principle, it did not matter whether the knowledge was true or false. As long as it was shared by the members of the society, it would hold them together. Moreover, in such a situation there was a necessity to protect it from criticism and to ensure that, if it was false, nobody could ever find out. For if people had found out that it was false, it would have ceased to do its job as a social bond, and the society in which the discovery that it was false had been made would have fallen apart. In addition to this kind of cognitive protectionism, there was also a presumption in favour of false knowledge. True knowledge makes a poor bonding principle, because sooner or later everybody will embrace it. It is essential for social bonds not only to bind some people together, but also to make sure that most people are excluded. One cannot, for long, form a society around the theory that the sun moves around the earth, because too many people would want to join and very few people would be excluded. But while there are very few true theories, there is an infinite number of false theories. Each particular falsehood divides the people who embrace it from all those who do not embrace it and who embrace a different kind of falsehood. By the use of false theories, one can divide mankind up into innumerable small societies. Hence there has always been a presumption in favour of the survival of false knowledge.

This situation puts knowledge at a real disadvantage. One might expect that false knowledge must eventually lead to the destruction of the people who hold it. But this expectation is proved wrong. While people with false knowledge about rain may fall behind in their food supplies, they will form such a solid solidarity that they can form better war-gangs than anybody else and thus take from other people the food which their false knowledge prevents them from growing. One could almost say that the very falsity of their knowledge enables them to compensate themselves for whatever disadvantages accrue to them because of false knowledge. The more false their knowledge, the greater the solidarity and the greater their ability to find such compensation.[4]

As a result of such misuse of knowledge, the acquisition of true knowledge has had a very slow and chequered career. The reasons for the non-cognitive use of knowledge are not difficult to understand. The misuse was dictated by a very simple necessity. We may assume that in the very first hominid days, societies reproduced themselves incestuously by the practice and prescription of marriage or sexual intercourse between brothers and sisters and first cousins. There may even have been no need for prescription. The sparsity of people cannot have offered much of an alternative. But when populations increased, it was found that such incestuous structures were condemned to remain small and, being small, were vulnerable. To enlarge societies, it became necessary to set up a taboo on incest and to ensure a certain degree of marrying out. But at the same time, it was necessary to limit the circle from which partners were to be chosen and to make sure that the social structure would not become diluted until everybody belonged and labour and food might have to be shared with the whole of mankind such as it then was. In other words, together with the development of an inclusion principle other than incest, there had to be a development of an effective exclusion principle. There had to be a circle of people who were in – people one could marry and eat with; and there had to be limits to that circle so that one could know which people could be preyed upon. The people who were outside the circle were simply part of the natural environment, like the leaves of trees, the roots of plants and the animals that were hunted. The great problem was how to define the extent of such circles. It is here that knowledge, beliefs and the rituals on which the knowledge and the beliefs were based came in handy. But false beliefs, which would distinguish one group from all other groups, were obviously more effective definitions of the circles than true beliefs.

The advantages of false knowledge had to be supplemented by another principle. The knowledge, though mostly false, had to be absolutely certain, or rather one had to be absolutely dogmatic in the belief that it was certain. Doubt could not be tolerated, and any form of criticism or scepticism amounted to a direct attack on the structure of society. Scrutineers and critics and proponents of alternative beliefs had to be eliminated because they struck at the roots of the social structure, not because they questioned the nature of the world. For this reason, the concept of certainty came to be associated intimately and inextricably with the concept of knowledge. Uncertain knowledge, it was thought, was not 'knowledge'.

Since in these early days people were not given to sociological thinking and were therefore not able to understand the real reason for the intimate connection in their minds between knowledge and certainty, they concluded, almost universally, that the certainty aspect of the knowledge they were holding derived from its *source.* If the source was right, the knowledge must be right. The source given was the teachings of an inspired prophet, direct revelation, tradition, established authority, Scriptures, or whatever. Eventually in Europe, during the seventeenth century, social structures loosened up and became more neutrally bonded; and more and more people started to question the pieces of knowledge which had been held and bandied around, and fewer and fewer people who did so were put to death and persecuted, because it came to be realised that one could change one's views on heliocentricity without endangering the division of labour and the cohesion of one's society. But strangely, or, perhaps, not so strangely, philosophers continued to believe that knowledge, to count as knowledge, would have to be certain knowledge and that certainty depended on the authoritativeness of the sources of knowledge. Thus they got hold of the wrong end of the stick. They rejected the traditional sources – revelation, tradition, authority and Scripture – and started to search for a new source which would be comparably certain and therefore yield comparable certainty. They found it at first in what they called a progressive refinement of common-sense 'observation': first in sensations, then in sense-data and finally, in our own century, in sensa and, eventually, in sentences which were alleged to be unadorned and ungarnished reports of sensations – the famous *Protokollsätze* of the Viennese. With the advent of looser and more neutral social bonding in western societies, they should have done it the other way round. They should have seen that 'certainty' was associated with the needs of social structure and had nothing to do with the condition of knowledge and that, given looser social structures, the notion of certainty could be dissociated from knowledge – a dissociation which would make any search for sources of knowledge superfluous.

However, it is no use crying over spilt milk. They had insufficient insight into the link between certainty and social structure and kept thinking that certainty was linked in an essential manner to the notion of knowledge. We can see today very clearly that the elimination of false knowledge and the growth of true knowledge – however it may take place – cannot be a question of the sources of knowledge. One could almost argue that this conclusion follows from pure logic. Anybody who thinks he or she knows what source would be the 'right' one – observation or sense-data or sensa or the unadorned linguistic expression of any of these or possibly the word of God in Hebrew or Arabic, as the case may be – *must,* in his or her heart of hearts, also think that he or she already knows what the world is like; that is, what he or she is proposing to find out. For the 'rightness' of a source can only be judged in terms of the nature of the world. If one does not know what the world is like, one cannot possibly know what the *right* method for finding out what it might be like would be.

If our ability to distinguish between what we take to be true knowledge from what we take to be false knowledge does not depend on the sources of knowledge, what does it depend on? Knowledge is a relationship between a knower and something that is known. The sources of knowledge would, of course, determine the nature of the relationship. But if it is not a question of sources, how is the relationship determined or set up? There are many possibilities of constructing models of such a cognitive relationship. One could suppose that the relationship consists in the way the world appears to a photon. Alternatively, one could suppose that the relationship consists in the way the world reacts upon the human mind; or, again, how it appears to a linguistic community. None of these possibilities are good models of the cognitive relationship. In all of them, one side of the relation is chosen arbitrarily. One might well ask why should there be cognition if one establishes the relation between the world and a photon? or between the world and a human mind? In all these cases in which one of the terms is chosen arbitrarily and could be replaced by any other arbitrarily chosen term, the relationship is not all that cognitive, because it really tells us more about the term we have chosen than about the world. If the world *appears* to a photon or to a mind, we can learn something about photons and minds; but not much about the world. If we consider knowledge to be the state in which something appears to something else, we imply that the knowing subject always detects qualities in the world it knows which would *not* be there but for the knowing subject: as in feeling, hearing, seeing, and so forth. In short, we are really looking for something other than an arbitrary knower whose knowledge can at best be an *appearance* of the world. We are looking for a knower who knows the world as it appears to itself or, in traditional terminology, how it is in itself. If one wishes to avoid the arbitrariness of such cognitive relationships, one gets involved in an infinite cognitive regress. In setting up the photon or the human mind as one of the terms, one could claim that the choice is not arbitrary, if one can show that the photon or the mind has some special quality or facility to 'know' the world and serve as one of the terms of the relationship. But such a claim can only be justified if one knows what photons or minds are like – that is, if one has prior knowledge and knows already what one is setting out to know. To have prior knowledge, one needs a relationship between a knower and a known and has to face again the old question as to what is to serve as one of the terms.

It is sometimes suggested that we should recognise that knowledge is a value and that the search for knowledge is like an ethical act of faith; and that the decision to look for rational knowledge is itself based on an irrational commitment.[5] In the end, this may have to be so; but before we commit ourselves to such a commitment, we should try to think of knowledge as a natural and especially a biological phenomenon – a value perhaps; but one that is embedded in our biological existence.

The most likely candidate for a model of the cognitive relationship, though this has been, as we shall see, far from obvious, is the model presented by biology. In this model we can see what counts as knowledge, how it is acquired, how it is retained and stored and what kind of information about the world is available. Biology provides us with a term of the relationship which is not chosen arbitrarily and the choice of which does not have to be justified by an argument which leads to an infinite cognitive regress.

We are here. In order to be here, we must have evolved. In the whole process of evolution from the Big Bang to the present moment, no sub-process is more complicated, more crucial and more telling than the first appearance of living cells in which DNA was related in a specific manner to proteins. Whatever the details of this sub-process, the appearance of the first living cell constituted a cognitive relationship between the cell and the world. The living cell is a special arrangement of molecules which is surrounded by a membrane which divides it from the world. The membrane protects the molecules on the inside in a very special way from the molecules on the outside. The membrane enables the molecules on the inside to remain in an ordered state because the cell sucks order from the outside world. In so sucking, it increases the disorder outside its membrane and protects the order inside its membrane. The membrane is therefore a highly significant and vital boundary between inside and outside. In order to perform its specific protective function, the membrane has to be selective as to what it allows into the cell. Some things have to be kept out and others have to be allowed in. The membrane, therefore, performs a cognitive function. It has knowledge about the outside world, and that knowledge makes it possible to discriminate between those parts of the outside world which can be allowed to cross the threshold and those parts which must not be allowed to cross the threshold.

> Now, what is distinctive about this cellular dynamics compared with any other collection of molecular transformations in natural processes? Interestingly, this cell metabolism produces components which make up the network of transformations that produced them. Some of these components form a boundary, a limit to this network of transformations. In morphological terms, the structure that makes this cleavage in space possible is called a membrane. Now, this membranous boundary is not a product of cell metabolism in the way that fabric is the product of a fabric-making machine. The reason is that this membrane not only limits the extension of the transformation network that produced its own components but it participates in this network. If it did not have this spatial arrangement, cell metabolism would disintegrate in a molecular mess that would spread out all over and would not constitute a discrete unity such as a cell.[6]

The living cell embodies knowledge about the world. It is the first and ontologically given candidate for the term we have been looking for in the cognitive relationship. Here we have a non-arbitrary model of the cognitive relationship, and since we may not presume that the cell is an arbitrary intruder into the world, we are entitled to think that the knowledge it embodies or represents is knowledge of how the world is in itself or how it appears to itself. At this point we can anticipate and add that that knowledge came to be lodged in the cell by chance mutation and selective retention, to use Donald T. Campbell's terminology. There must have been myriads of early cells which embodied wrong information about the world and could, therefore, not suck in the correct parts of the world. They did not survive, at least, not long enough to divide. The cells or cells which happened to embody the correct knowledge survived, divided and multiplied. The transfer of information could be called a process of learning. Such learning did not result from observation of the world and a storage of the information which pounded repeatedly and long enough upon the cell's membrane. There was nothing in the cell to enable it to learn from instruction and observation. For that matter, the world outside the membrane did not consist of discrete bits of information which could have been picked up by careful observation. The outside world, to use Gerald M. Edelman's phrase, 'is not labelled'. Any labelling we find today is the result of the cell's response to the world and of the way the knowledge it embodies makes it react. Fred I. Dretske is quite wrong in thinking that 'the raw material is information'.[7]

The fact that the outside world does not consist of discrete bits of information which can be processed by the cell the way a computer processes information is crucial. Technically, one speaks of information processing when output is a reliable indicator of input or the other way round. Since the world does not consist of discrete pieces of information, the establishment of a cognitive relationship between knower and known cannot be likened to technical information transfer from the to be known to the knower. Moreover, the knower does not simply store pieces of information. The knower – in what can be properly described as a cognitive and non-causal relationship – possesses the ability to categorise and abstract and thus create a picture of it or his or her world through these abstractions and to respond to it in terms of the results of such abstractions. This cognitive relationship is very different from information processing. If in the following pages there is talk of 'information' and of 'information storage' and 'information transfer', the term 'information' is to be understood in a non-technical sense and not in the sense in which it is understood by people who deal with computers and keep hoping for artificial intelligence. 'Information', throughout, simply means pieces of knowledge; and 'information transfer' should be taken as nothing more technical than 'knowledge acquisition'. We speak of 'having information' and take this expression to be synonymous with 'having knowledge'. Under no circumstances should it be read as meaning that the world contains information which can be transferred to something like a computer and be processed.

The whole of evolution, from single cells to *homo sapiens*, including all the branchings out which have taken place, is to be seen as a process of knowledge acquisition. The organisms are one term of the cognitive relationship, and the environment, the other.[8] As organisms evolved away from the primitive state of the first single cells, there eventually emerged organisms with consciousness. Consciousness must be seen as a function of a nervous system above a certain degree of complexity. There are probably degrees of consciousness, depending on the degrees of complexity of the nervous system of the organisms involved. It is quite impossible, so far, to determine what the minimal complexity for the presence of consciousness is. As Gerald Edelman put it, we have to accept the following sequence: world (matter), then earth, then living forms and evolution, then animals, then primates and hominids, then social communication and language and culture and, finally, science.[9] Somewhere along this sequence, at a point we do not really know of and certainly after *x* and most probably before *y*, we encounter the phenomenon of consciousness.[10] After an initial confusion which identified it as *res cogitans*, philosophers as well as scientists have tried to shy away from it. It has been variously defined and is usually regarded as a dangerous concept, even though the phenomenon itself is undeniably present. Whichever way, it is mysterious in its presence, and the reasons for its evolution are not understood.

It is even often believed that the quality of our knowledge depends on our ability to ignore the phenomenon, and it has frequently been pointed out that one can do a lot of physics, psychology and neuroscience without worrying about consciousness. 'In most psychological modelling of human cognition, no mention is made of consciousness or intention.'[11] This is perfectly true. But the fact is that there *is* consciousness, and any physics, psychology or neuroscience which ignores it must be incomplete and unsatisfactory. Though the view that the more consciousness is ignored, the better the knowledge is patently absurd, it is amazingly popular. The popularity of this view is, of course, somehow linked to the superstition that all mention of consciousness is an intrusion of subjectivity and a departure from objectivity.

Karl Popper remarked that a theory of the non-existence of consciousness cannot be taken more seriously than a theory about the non-existence of matter.[12] If consciousness is there, it must have evolved; and if evolved, there must have been a reason for its evolution – that is, there must be an adaptive advantage in having it. 'The fact that consciousness has been evolved', writes W. H. Thorpe, 'is itself strong presumptive evidence that it is not an *accidental* by-product. It suggests that at least this aspect of mental life can accomplish something for which a neuronal mechanism alone, however complex and elaborate, is inadequate.'[13]

If it is bad enough that there has been serious doubt whether the extrusion or elimination of consciousness from knowledge might not be possible and advantageous, it is even worse that we are unable to pin-point consciousness. As William James put it long ago, consciousness is something 'we know so long as no one asks us to define it'.[14] There has nevertheless – and this is hardly surprising, given the pervasiveness of the phenomenon and its importance – been a plethora of definitions and attempts to describe it. Consciousness, it is often asserted, is intentional. It is *about* things and always has a content. Brentano's famous old formula has recently been slightly recast by Ted Honderich: 'Consciousness consists in general of [the] interdependent existence of subject and content.'[15] Honderich goes on to say that this interdependence is such that 'there can be no subject without a content, and no contents not in relation to a subject'. More esoterically, Roger Penrose believes 'consciousness to be closely associated with the sensing of necessary truths'. D. R. Griffin has been one of the few investigators to consider seriously that animals other than man are capable of consciousness.[17] He believes that all that is required for consciousness is that there should be 'mental images'. Whatever the definition, Sperry has called it the 'Number 1 problem in brain research'.[18]

The weird mystery of the whole phenomenon has often been remarked on by philosophers. Biologists and ethologists are equally mystified. Even more weird than the phenomenon itself is the grammatical structure of the sentences which contain the word. We always say we are conscious *of* something or *that* something is happening. But we cannot say that we are just conscious. When we do, we do not mean that we are just conscious, but merely that we are not in a coma or not asleep. At the same time, it is known that we cannot be conscious of the neuronal processes which make us conscious, and this fact has removed talk of consciousness and mind to a mysterious realm of spirituality and even given rise to the surmise that consciousness may be a separate substance. It is less well known that we cannot say with any degree of assurance what precisely it is, at any one moment, we are conscious of. Philosophers and psychologists usually jump to the conclusion that we always know with absolute certainty what we are conscious of – a chair, a hope, a pain; and, what is more, not only that such conscious states are certain to the person who has them but also that they are incorrigible – certainly incorrigible by the person who has them. It is of course granted that we may have no rational or perceptible reason for hope or for being conscious of a chair or of whatever; but it is universally assumed that we are incontrovertibly certain of the fact that we are conscious of hope, pain or chair.

The entire discussion about the reason why there is consciousness is bedevilled by these conditions. Unless consciousness were an adaptive advantage, it would not have evolved. So there *must* be an adaptive advantage in having it. This is not to suggest, as radical adaptationists often do, that the presence of consciousness is to be explained by the selective advantage it confers as it stands. On the contrary, most probably, consciousness is an accidental by-product which, like the spandrels of San Marco,[19] owes its retention to the uses other features made of it. On the face of it consciousness does not appear to be advantageous at all. Unless one fallaciously presumes – as I shall argue presently and again at length in Chapter 1 – that every state of consciousness is inextricably linked to an articulated linguistic expression and, therefore, known or knowable, consciousness is little more than a power for mischief. When it appears in as strong a manner as it appears in human beings, it is inchoate, difficult to pin-point, undefinable, with qualities which are elusive and which escape us the harder we try to seize them: consciousness has a powerful potential for mischief because it can make its possessor relentlessly at odds with the facts of life. Biologically, the human animal, like any other living being, is part of the long chain which has evolved from the beginning of life in the very distant past. Every single individual living being is a link in that chain and a passing stage in the continuity of living cells. For all developed living beings, it takes two to produce one, and that one must again associate with another to ensure the continuation of its cells. Single individuality is more an appearance than a reality, or, as in Buddhism, a delusion rather than a reality.

In this situation, consciousness provides an opportunity for diversion and a possibility of directing attention away from the reality of living matter. The very uncertainty of consciousness, as will be argued presently, is a pressure which makes for the evolution of a three-dimensional language. And with a three-dimensional language there comes the possibility of identifying any one set of states of consciousness as an *ego* or of interpreting it as belonging to a special person. Via language, consciousness therefore encourages the illusion that the biological organism, provided its neuronal system is powerful enough to generate strong consciousness, is at cross-purposes with the transitory nature of its biological existence, for it encourages the belief that the organism has a special reason for existing and that its existence ought not to be temporary. Since it potentially confuses the issue, it may look at first sight as if consciousness is not an advantage at all. But this cannot be right; for if it were not an advantage, it would not have evolved. We must, therefore, take a second look.

In plain and flagrant defiance of evolutionary biology, which shows us that consciousness would not have been retained if it had not been adaptive and had not given organisms which had it an advantage over organisms without it, there is an abundance of philosophical opinion that consciousness has no advantages and that we can do without it. Rorty, for example, believes that 'consciousness' is just a word we use to refer to certain regions of public space and time like brains and behaviour and the relations between them.[20] Others invoke Freud to support the view that consciousness has no explanatory value,[21] and according to H. L. Dreyfus,[22] Heidegger maintained that consciousness, separating out subject and object, is neither necessary nor sufficient for human activity.

It has to be admitted that the search for the nature of the advantage of consciousness has been muddied by the assumptions (a) that we are always conscious *of* something; and (b) that we know with precision what that something is. On these two related assumptions, everybody keeps searching for that advantage in the wrong direction and keeps wondering why it should be an advantage to be *conscious* of pain or of a chair rather than just reflexively respond to the pain or to the chair. There are philosophers who argue that since there is no advantage in having doubles (i.e. chair + consciousness of chair), the presence of the faculty that makes us see double must for ever remain a mystery. Philosophers like Nietzsche, equally perplexed, believe that the faculty which makes us see double is mischievous, will confuse and inhibits behaviour to a point at which it becomes the cause of our destruction. The more correct and promising way is to introspect more carefully and more attentively and to free ourselves from the constraints of our grammatical rules which oblige us to speak always of being conscious *of* something, so that we can discover, in the first instance, that when we are conscious, we are not conscious of anything very specific and that the initial stirrings of consciousness are, to start with, inchoate. There is good reason why we should make the discovery that in the first instance these stirrings are silent. The reason for this discovery is that at long last we are beginning to take neuroscience very seriously, and this means we are considering the consequences of its findings for our philosophical understanding. According to these findings, consciousness must be a neural phenomenon, and for that reason it must be, in the first round, as silent and unworded as neural events themselves. 'Sense-data', J. L. Austin used to say, 'are dumb.' Unfortunately he did not add that, therefore, the term 'data' ought not to be used.

It is almost ironical that one of the most important contributions of neuroscience to philosophy should be a negative one.[23] Neuroscience makes us see that words and language cannot come from neuronal, unworded events *without* an intervening development. We must take it that the intervening development is the stirrings of consciousness. Since that consciousness is directly linked to neuronal events or, rather, to the complexity of the totality of these events, it too must, in the first instance, come without words. It must be pre-linguistic and cannot, as it emerges, be identifiable in terms of worded labels.

Once this is accepted, one will see that consciousness of a chair or of pain is not a duplication of an event we are conscious of; but, more correctly, a hypothetical interpretation of the fact that we are conscious. Such a conclusion would force us to revise the grammar of sentences in which the word occurs, so that we can begin to speak of consciousness of nothing in particular and then regard any specification of such inchoate consciousness as a hypothetical interpretation.

The shortcomings of all attempts made so far to pin-point the phenomenon of consciousness are that they inextricably link consciousness either with language or with specific mental images – which, closely examined, also derive from language's ability to specify distinct objects or events. These attempts do not examine the link between language and consciousness. They simply take it for granted. The reason why this link cannot be taken for granted and, in the end, has to be rejected is that it is most probable that some animals which are not capable of language have a nervous system sufficiently complex to allow some stirrings of consciousness. Any attempt to define consciousness as a relation between a subject and a specific content requires and presupposes linguistic competence. For this reason it is highly questionable.

I shall call the view that conscious states are essentially specific and specifically knowable 'the slapdash assumption', and postpone a fuller discussion of this assumption to the following chapter. It is an assumption which is widely made, and it is easy to see why it is widely made. After all, as the word 'conscious' indicates, according to both semantics and the grammatical rules of western languages, when we are conscious, we are conscious *of* something; and it is therefore natural that we should think of consciousness always in conjunction with the object or event or feeling we are conscious of and that, moreover, we should take this conjunction to be hard and fast so that we cannot extricate ourselves from it and that, indeed, to do so would force us to go against the very rules of our language. The sensible inference would be that we ought to be more wary of the slings and webs our linguistic usage is winding around us. But, unfortunately, most philosophers have seen their way to taking it the other way round. They have always preferred to conclude that since our language does not permit a use which would tell of inchoate consciousness not linked to words and therefore essentially unidentifiable, thoughts of such consciousness are illicit or unfounded or both.

The assumption is widespread. Let us look at two philosophers as diverse as Fodor and Quine. *Quidquid delirant reges plectunctur Achivi*! If the philosophers are content with the assumption, one can hardly be surprised to find almost all neuroscientists equally content. I will postpone discussion of the consequences of the assumption for neuroscience to the next chapter and confine myself here to a few remarks on the philosophers. Start with Quine: 'What we need,' he writes,[24] 'as initial links ... are some sentences that are directly and firmly associated with our stimulations ... The sentence should command the subject's assent or dissent outright, on the occasion of a stimulation in the appropriate range, without further investigation and independently of what he may have been engaged in at the time.' And take Fodor, who observes, correctly, that there cannot be a neural event corresponding to or causing the enunciation 'dog'; for if there were, the same event would have to occur when we intend 'dogma' rather than the canine animal. He concludes that, therefore, a mental intention must inter- or supervene. So far, so good. But since Fodor clings to the slapdash assumption that there cannot be anything mental unless it is inextricably glued to a worded label, he forces himself to believe that there must be a language of thought – Mentalese. He does not realise that the moment we let go of the slapdash assumption, we can entertain the possibility that speech and speech rules originate somewhere else and merely happen to come together in a hypothetical manner when we say we believe or desire or fear or assert this and that. Without the slapdash assumption, there is no need to imagine that there is a language of thought. In Fodor's view, language and mental event are inextricably connected: 'You can deduce', he writes,[25] 'the causal role of a mental state from the semantic relations of its propositional object.'[26]

If more detailed discussion of the slapdash assumption and the consequences of slapdashness will be postponed to the next chapter, it is important here to indicate that some well-known philosophical controversies thrive on this slapdash assumption and that, since the assumption *is* slapdash, there is good reason for steering clear of them. If one assumes that all conscious states are identifiable in so many words – as pain, as hope, as awareness of a chair, as anticipation of another person's action, and so forth – one will start wondering whether there is such a thing as introspection, as a private language, as the incorrigibility of statements about feelings and whether feeling-talk is merely a duplicate of physical language or whether mental feelings are substantially distinct from neuronal events. Arguments about these questions fall into the category of questions which Kant called antinomies. By this he meant that one can argue with equal plausibility for both sides. He was thinking of questions about the infinity of the world and was able to show that for reasons which need not concern us here, the arguments for and against infinity are equally convincing. I would like to show that the argument for introspection, for a private language, for the incorrigibility of statements about inner feelings, and for the eliminative materialism which states that feeling-talk is merely a duplicate of talk about neuronal events is as plausible as the argument against all these beliefs. The fact that these questions lead to antinomies which cannot be resolved stems directly from the slapdash assumption that in all cases states of consciousness are shot through with words and are clearly identifiable.

Consider, for example, the question of introspection. If one makes the slapdash assumption, one can argue that introspection is possible because one simply knows, by definition, what one is conscious of. But one can argue with equal plausibility that introspection is not possible or does not lead anywhere, because one may not be conscious of what one is *really* feeling or thinking, and that, since introspection equates awareness with knowledge of awareness, one cannot do a double check on what one thinks one knows by introspection. Or take the endlessly debated problem of eliminative materialism. One can argue plausibly that talk about feelings or mental events is genuinely different from talk about allegedly corresponding neuronal events because one would not engage in it if talk about neuronal events could convey the same message or meaning either to oneself or to others. One can also argue plausibly to the contrary and say that since there can be no mental-spiritual substance, feelings and other mental events must be identical with neuronal events. Or again, consider the arguments about a private language. On one side it is maintained that there can be no such thing as a private language which bestows meaning on states of mind, because sentences derive their meaning from their use according to rules; and rule following is not possible if one privately invents sounds and uses them in an unruly way. On the other side it is argued that meaning is stamped upon publicly intelligible speech by inner intentions formulated as a language of thought – a species of Mentalese. When expressions are established in Mentalese, they can then be translated into a used public language. Without prior Mentalese, it is argued, no meanings can attach themselves to sounds, whatever their order.

There is no need to rehearse more antinomies.[27] But one should note that in all these cases, the fact that these questions lead to antinomies derives from the slapdash assumption that every state of mind is inextricably linked to a set of words. As soon as one is prepared to admit that there are non-worded, pre-linguistic neuronal events which are sufficiently complex to generate consciousness without words – consciousness which is inchoate and unidentifiable – and that the labelling of states of consciousness is a *subsequent* or supervening event, these questions can be answered in a more unequivocal way and the plausibility of some of these arguments will disappear or, at least, diminish. Further discussion of the implications of the slapdash assumption will be found in the following chapter.

The problem of the definition or description of the phenomenon of consciousness is closely connected with the problem as to why it has evolved and been selected for survival. Its emergence may have been a fluke; its survival cannot have been. One must therefore ask what precisely the adaptive advantages of consciousness are. Richard Dawkins wrote that the evolution of subjective consciousness is to him 'the most profound mystery facing modern biology'.[28] All answers I have ever come across have been bedevilled by a traditional psychology which derives from Plato. Plato believed that the soul – or consciousness, as we would now call it – is the director of human behaviour, capable of monitoring, as we nowadays would put it, its own discriminational capacity. The more it directs or monitors, the better the behaviour. This belief is superficially plausible as well as attractive and has gained a very firm hold. It was only at the time of the Romantic revolution that thinkers from Rousseau and Hume onwards started to have their doubts about the value of conscious direction. They had a good forerunner in Shakespeare's insight that 'thought is the slave of life' and were followed up by Nietzsche and, eventually, by Sartre's insistence that existence precedes essence, i.e. that we define ourselves not by conscious reason, but only *after* we have acted and as a consequence and in terms of such action. But for the most part, the Romantics got hold of the wrong end of the stick. Having doubted the benevolence and beneficiality of consciousness and reason, they saw it as a hindrance and as a destructive faculty. In the Romantic tradition, consciousness was a factor opposed to the forces of life, inimical to intuition, to the heart's desire and to biological appetite. Nietzsche put it in a nutshell when he proclaimed that with consciousness and reason mankind had prepared for itself a good instrument of self-destruction. The truth, I will argue, is infinitely more complex and infinitely more interesting.

If one follows the Platonic tradition, it is easy to see why consciousness is an adaptive advantage. It promotes intelligent behaviour and permits the development of theories which anticipate the future, and by these means simply increases the survival chances of conscious beings. This is what G. M. Edelman's list of the adaptive functions of consciousness more or less amounts to,[29] and almost all other writers on the subject one cares to consult would agree with him. N. Humphrey, for example, in his *Consciousness Regained*,[30] explains that in human societies it is an advantage to be able to anticipate and imagine how other members of the society are likely to react and argues that such anticipation and imagining are due to consciousness. Hence, consciousness is an adaptive advantage and has been selected for. Alternately, it is believed that animals capable of consciousness have a better way of maximising inclusive fitness than animals without consciousness.[31]

The difficulty with these and similar reasons for the adaptive advantages of consciousness is that they assume that in all cases consciousness is always, and always immediately, linguistically expressed. This does not seem to be the case, and as long as we fail to distinguish between raw consciousness and linguistically expressed consciousness, we will never be able to arrive at a proper understanding of the reasons why consciousness is an adaptive advantage. Raw or unworded and uninterpreted consciousness is prior to language[32] and, indeed, as will be argued shortly, is the driving force for the evolution of a special three-dimensional form of language, though not of purely communicative language. When we reflect by introspection – and that is all we can go by to improve a little on William James' justified doubts – one can see that consciousness is prior to language and manifests itself in some pre-human animals who are incapable of language. Whatever the truth about pre-human animals, human beings, when they introspect carefully, realise that Nietzsche was correct in saying that a thought comes when 'it' wants, and when I think, I am not entitled to say that it is 'I' who am thinking until I have committed myself to a linguistic convention. The identification of a feeling or a thought as stemming from an ego is an *interpretation* of the feeling or the thought. Primarily, all we can speak of is that 'it' has a feeling or a thought.[33] In any case, if the Plato-derived reason for the adaptiveness of consciousness is to be upheld, one commits oneself to the assumption that language attaches itself inextricably without further ado to consciousness or that language and consciousness emerge and are practised simultaneously and in tandem and that if there is consciousness in pre-human animals without language, there must be some quite different reason why it was adaptive for them. The stand of the Romantics is even more difficult. If consciousness is an impediment, it is impossible to explain why it was selected and retained. It could well have arisen somewhere as a chance mutation or a by-product. But unless it was an adaptive advantage, it could not have been retained, and certainly not retained on such a massively widespread scale.

I will propose a completely different theory of the functioning of consciousness in its relation to language and the reasons for its selection and adaptive retention. The theory is based upon the consideration that animals – and I am here including the human animal – can manage perfectly well without consciousness. Goal-directed behaviour can be produced by neural systems of various levels of complexity without consciousness and without conscious representation of the goal or of the stimulus which triggers the behaviour towards the goal. There are self-organising neural processes which achieve an end-state of interaction between the organism and its environment in a flexible and adaptive manner.[34] If there is a modicum of consciousness in many non-human animals, it goes unexpressed, and we must think of it as a by-product of their nervous system which does no harm and which therefore has never been selected against. If such consciousness as is present in felines or apes had ever caused inhibitions and prevented them from goal-directed behaviour by making them 'think' without words and causing them to show a certain scepticism towards their instinctive and programmed responses, it would have been eliminated as a non-advantageous factor long ago. Consciousness without language is neither here nor there, although it undoubtedly exists.

The case of the human animal is different. In human beings, consciousness is initially inchoate and remains so until it is given a local habitation and a name by language. Such consciousness as is probably present in some higher animals is equally inchoate, but cannot be given a local habitation and a name. But once it is linked to language, it plays an entirely novel role. To start with, it produces inhibitions and scepticism as to triggered responses and delays action and goal-directed behaviour. Such language-linked consciousness makes the human animal less adaptive, not more adaptive. The question then arises why it has been selected for and retained in spite of the fact that initially it retards responses and appears to be a disadvantage. The answer to this question is to be found in the peculiar way in which it is linked to language. Inchoate consciousness, I propose to argue, is a consciousness which we share with many other animals; but in human beings it becomes the *cause* of the evolution of a specifically human form of language which is qualitatively different from all other pre-human forms of communication. There are two reasons why consciousness in humans – though it is impossible at present to determine the exact point of hominid evolution at which this took place – plays this role. The first is that in human beings, because of the complexity of the nervous system, consciousness is so potent that it becomes inchoately aware of its inchoateness. Cats and dogs are most probably conscious, inchoately so. But there is nothing in their behaviour to indicate that they are in any way aware that their consciousness is inchoate and that with a bit of struggle and pressure it could be given a definition. The goal direction of their behaviour is, on the contrary, taking place *in spite* of their consciousness. In human beings, the awareness of inchoateness creates an unease and discomfort which exercises a pressure towards definition and articulation. The second reason is that in human beings, possibly only after the Neanderthal period, there was a vocal tract and a brain area which permitted, under the pressure exercised by the unease of the awareness of inchoate consciousness, the development of a wide range of distinct sounds and of syntactical rules to govern their emission.

It is important to understand this proposal correctly. Language as a means of communication of matters of fact and of messages is widespread among animals. Such two-dimensional language is as old as life itself. 'All organisms', writes J. Z. Young, 'plants and bacteria as well as animals, have such communication channels, through which they send messages in code ... A code is a set of physical changes that are used in various combinations to evoke specific responses by a receiving agent that is tuned in to receive them.'[35] Bird song, the whistling of whales, bee-language and the gestures and cries and body language of many animals are able to do this. Such languages, remarkable as they are, are confined to communication of existing states of affairs. The messages they convey are communications of how things are – where there is food or whether there is a mate or whether there is danger. For this reason Jonathan Bennett has argued,[36] incomprehensibly, that bee language is not a language. Two-dimensional language can even be stretched to deceive a predator; but even in this case, the communication is a plain statement of an existing state of affairs. The lie which is communicated is not a free invention but an attempt to hide something which exists and which is the case. There are birds which give a warning cry when there is *no* hawk. The other birds fly away and the bird which gave the 'deceitful' cry gets all the food. Dawkins has wondered whether such ability to deceive is the beginning of consciousness.[37] But it is important to note that this kind of deceit can be explained as an instance of two-dimensional language. For the 'deceiving' bird does not say to itself: 'I will pretend there is a hawk.' The deceiving bird, on the contrary, has evolved the ability to give a warning cry when there is *no* hawk, because it was clever enough to notice that when birds fly away when the cry is given, it can eat all the food there is around by itself. The bird is not 'deceiving'; it is giving the warning cry as a message about an existing state of affairs: when the other birds depart, there is more food for it.

People who have tried to explain the evolution of language have always taken it for granted that human language is simply a sophisticated continuation and extension of such earlier languages, and they have regarded the whole problem of the evolution of language as if it were a problem which would have to be solved by considering human language as a mere extension of animal language and gesturing. It is highly improbable that specifically human language should be a mere extension of earlier communicative gesturing, and we will indeed simplify the problem if we take communicative gesturing for granted – whatever its origins – and explain the evolution of *human* language only as from the moment at which potent consciousness had emerged. We need not interest ourselves in the evolution of message-giving ability; but only in the evolution of human language which allows phantasising, speculation and descriptions of events which do *not* take place. Even the displacement and the scenario building which involves an ability to communicate information about the future and about distant places, an ability which is certainly present in bees[38] and possibly in some primates,[39] is a form of communication of matters of fact and therefore qualitatively different from what I shall call three-dimensional language. Three-dimensional language can, therefore, not have evolved as a simple extension of two-dimensional language. To show that there is continuity and that three-dimensional language is nothing but an extension of two-dimensional language, one would have to show, for example, that bees can pass on information about places which do *not* contain honey or even about places which do *not* exist.

It is important to understand the relationship between two-dimensional and three-dimensional language correctly. Most writers on language make no absolute distinction and simply take it that eventually, given the development of a suitable vocal tract and of larger brain areas, language became more explicit, vocally flexible and more plastic. The view taken here is very different. Two-dimensional language, whatever its evolutionary origins, was merely the raw material from which three-dimensional language, given the pressures of inchoate consciousness, evolved. The process is not to be compared to gradual training which increases muscle power; but to the process which led from gills to lungs. Ability to use oxygen as a source of energy was the raw material which constrained subsequent evolution along a certain path and precluded other possibilities. But lungs are not simple extensions of gills. They evolved to use oxygen in a totally different environment and under very different pressures. In human beings, the ability to use language two-dimensionally for simple communication of messages about matters of fact is more like a fossil, like the rudimentary or atrophied gills in a lung-breathing animal. The origin of two-dimensional language may well go back, as Darwin surmised, to the principles of semantic efficacy.[40] Communication of meaning by expressive gesture was the spin-off of independent forms of natural behaviour. The spreading of wings, a natural precondition of flight, may come to signal aggression. Or, as Pavlov showed, 'anything confusedly connected with the physical *stimulus* of a natural reflex action is liable to become a *sign* which triggers the same reflex response.'[41] But this problem does not concern us here. We are interested in what happened *after* the evolution of two-dimensional language.

I would like to propose that human language is fundamentally different and that the explanation of the origin of human language must start somewhere else. Languages which are confined to the description and communication of matters of fact, including certain types of pseudo-lies, are two-dimensional. By contrast, human language is three-dimensional and qualitatively different from two-dimensional languages. By a three-dimensional language I mean a language which has optatives, conjunctives, future and past tenses and which has a sufficiently flexible syntax to express representations, images, wishes, hopes, regrets, and so forth, of events which have not taken place, which may never take place or which have not yet taken place. In such a three-dimensional language one can state inventions and imaginings far beyond anything suggested by an observation of what there is. One can state a type of lie which is in principle quite different from the self-serving lies of which some two-dimensional languages are capable. Popper in his *Objective Knowledge,*[42] though he does not use the terms 'two-dimensional' and 'three-dimensional', explicitly distinguishes between these two kinds of language, but does not go into the question as to what the evolutionary pressures, which forced the development of three-dimensional language, are. Above all, one must bear in mind that the origins of a three-dimensional language, unlike the origins of a two-dimensional language, must not be sought in the need of communication. A three-dimensional language, it is true, can, among other things, also be used for plain communication. But this is merely an interesting and advantageous side-product. It had its origins in other needs which had nothing to do with the communication of messages about matters of fact. Plato, I think, was one of the first to have observed that human speech exists, in the first instance, not for communication, but for something like an interior monologue – a way in which the soul speaks to itself.[43]

Let us go back to the initial inchoateness of all consciousness, human and pre-human. Provided the nervous system which produces it is sufficiently complex, such consciousness will have a certain colour or quality – it will consist of 'qualia'. These qualia cannot be very distinct or defined, but, broadly, one might distinguish happy ones from unhappy ones, terror from peace, and so forth. A more detailed discussion of such inchoateness and its consequences must be postponed to the next chapter. All we need here is an understanding of the difficulty of pinning conscious states down. One realises that any state of mind infinitely transcends the savours or the chair or any conceivable experience or sensum which one thinks might have 'caused' it. It becomes plain that the object of one's quest does not lie in any of these so-called causes, but in oneself. Any so-called cause calls up in one, but does not itself explain, and can only repeat indefinitely, with a gradual loss of strength, the same testimony; which one cannot interpret, though one hopes at least to call upon it again and to find it there intact and at one's disposal, for one's final enlightenment. And then one lets go of the so-called cause and examines one's own mind. It is for it to discover the truth. But how? What an *abyss of uncertainty* whenever the mind feels that some part of it has strayed beyond its own borders; when it, the seeker, is at once the dark region through which it must go on seeking, where all its equipment will avail it nothing. Seek? More than that: create. It is face to face with something which does not, so far, exist, to which it alone can give reality and substance, which it alone can bring into the light of day.[44]

Such inchoate consciousness, struggling with itself to define borders it surmises it might have, will produce a certain unease and discomfort. It will make the organism in which it arises a little anxious and uncertain – a state comparable, perhaps, to the state when something is on the tip of one's tongue, but one cannot think of the right word; or a state one is in when one cannot remember something one thinks one knows, a state, possibly similar to the state of *déjà vu*, where the more one tries to focus on the earlier thing one feels one has seen, the more distant and foggy it becomes. In this condition of inchoate consciousness the two-dimensional language we must presume our very distant ancestors were capable of and the origins of which need not concern us here came to be transformed gradually into a three-dimensional language. The evolutionary pressure consisted in the fact that qualia cannot be described, because they are airy nothings, not a *res cogitans*. In order to articulate qualia and indicate and define them, one needs a language capable of three dimensions; that is, capable of inventing hypotheses and of constructing symbols by shuffling and reshuffling matters of fact so that the imagination can body forth the forms of things unknown which can be turned into shapes. If consciousness were a *res cogitans*, two-dimensional language would have been sufficient to make descriptive statements about it and formulate messages to other con-specifics about it. The peculiar non-substantial quality of consciousness was the pressure which transformed language from two dimensions into three dimensions. Such a transformation presupposed that the two-dimensional language was vocal and the vocality sufficiently flexible to allow amplification of distinct sounds. In human beings, we know, a prior evolution of the ability to make such sounds had taken place. There had been sufficient enlargement of the brain area and a development of a vocal tract through the supralaryngeal space to allow for the production of sounds much more diverse and distinct than, say, the sounds produced by birds or whales. When these two products of evolution – a suitable vocal tract and a sufficiently complex neural system – came to coincide, there began the transformation of two-dimensional language into three-dimensional language.

Two-dimensional language is not specific to humans and is certainly not man's primary attribute. If one is looking for man's primary attributes in this direction, one should say that it is the combination of consciousness with a sufficiently suitable vocal tract and brain area – a combination which made for the transformation of two-dimensional language into three-dimensional language. Consciousness by itself is not much of an adaptive advantage.

Though it exists, it also exists in many higher animals in which it does not fulfil a specific purpose. Like higher animals, human beings could conceivably have done without it. Though it was an accidental by-product in higher animals, it would not have been selected for in human beings had it not been for its ability, in combination with the suitable vocal tract and the necessary brain areas, to press for three-dimensional language. Three-dimensional language, whatever else it can do, also makes it possible for consciousness to become articulate and articulated. For, since consciousness is not an existing substance, a *res cogitans*, the articulation of consciousness is a matter not of descriptive reference, but of hypothetical vocabulation and symbolisation. Consciousness, not being a substance with qualities, is not something one can describe. That is, it requires a language which can do more than describe existing states of affairs and which is capable of indicating *non-existing* states of affairs. Consciousness, in this argument, is not seen as designing language, but as an evolutionary pressure making for the development of a specifically non-communicative, non-descriptive, three-dimensional language, which does something a two-dimensional language, derived from gesturing and from conditioned reflexes without the intervention of consciousness, cannot do. To avoid misunderstanding, one must stress that the argument that consciousness is an evolutionary pressure should not be confused with or mistaken for the notion that language is consciously designed or designed by consciousness. Potent, non-articulated and inchoate consciousness has to be prior to three-dimensional language, though not prior to two-dimensional language. This priority cannot be detected and is in fact usually denied[45] by all those people who make the slapdash assumption that consciousness and language-articulated consciousness always go together. On the slapdash assumption, one cannot see that consciousness is adaptive *because* it acts as a pressure and *because* it makes for the evolution of three-dimensional language. If the slapdash assumption were justified, one would have to think up a completely different reason for the adaptiveness of consciousness.

A three-dimensional language has enormous adaptive advantages. Inchoate consciousness, provided it is as strong as it is in human beings, is not only unnecessary as it stands, but a positive disadvantage. It delays action, causes hesitation and inhibition and makes cooperation of individuals difficult. It is because of the strength of consciousness and the strength of its inchoateness that man, of all animals, gets into social difficulties. 'The social complications', writes Larry Weiskrantz, 'are a consequence of the powers of consciousness rather than the other way around.'[46] Such consciousness must be seen to have been an evolutionary overkill. At first, the nervous system became more powerful and more complex than necessary. As far back as 1869 A. R. Wallace had wondered that natural selection could have produced a brain only a little superior to that of the ape, whereas man 'actually possesses one but very little inferior to that of the average members of our learned societies'.[47] Similarly Darwin wondered why the brain should be so large and doubted whether this could be due to *natural* selection.[48] Given the seeming disadvantage of inchoate consciousness, we must regard the evolution of three-dimensional language, in the first instance, as some kind of damage control. In its ability to provide labels for these inchoate states of consciousness, three-dimensional language not only limited the damage but, as we shall see shortly, turned the damage into an asset.

Three-dimensional language, like two-dimensional language, can be used for plain communication of messages about matters of fact. But here it is actually less useful and, supposedly therefore, less adaptive than the language of bees. For a three-dimensional language is capable of endless rhetoric and makes the plain communication of straight matters of fact usually quite difficult, for even simple messages about what is happening are frequently misunderstood when expressed in a three-dimensional language. For this reason, Rudolf Carnap tried to prune our three-dimensional language of its potential for rhetoric and formulate rules for translating statements made in three-dimensional language into 'equivalent' (*sic*!) two-dimensional language statements. He obviously must have been of the opinion that bees and birds and whales have it all over us! Contrary to Carnap and his school, the communicative capabilities of three-dimensional language must be regarded as something of a fossil which was left over from the period in which pre-human beings used nothing but two-dimensional language. Leaving aside the capacity of communication and its difficulties, three-dimensional language has an enormous adaptive advantage precisely because it can make statements about events which have *not* taken place, about events which do not and cannot exist.[49] It can distinguish events of the past from events of the present and the future. It can express wishes, hopes, aspirations, ideals and fears and, what is more, even unreasonable and unjustified fears as well as phantasies and images of events which have never been known to happen. With such potential, plain communication of what is actual can indeed be seen to be difficult. But the ease of invention of three-dimensional language is crucial, for it is here that its advantage lies. With a three-dimensional language one can formulate, for example, statements about regularities and generalities which can never be observed to take place – for all one can ever observe is a very limited number of instances. With the help of three-dimensional language one can formulate images about structures which cannot be seen, even though they might exist. One can carry out a *Gedankenexperiment,* one can argue either with oneself or with others and one can entertain hypotheses. Three-dimensional language, in short, makes possible a vast extension of knowledge and increases the potential for the acquisition of knowledge through hypotheses and the testing of hypotheses and the elimination of false hypotheses.

The most important aspect of having a three-dimensional language is that it enables the organism to go beyond the information given. This can be done either through inventing events which have not taken place or through making mistakes about events which *have* taken place. It is, in the last analysis, the ability to make mistakes which provides the most fruitful repertoire for gaining information about the world. If there is an abundance of mistaken judgements and theories, one can select and eliminate false judgements and theories as errors. This method, as we shall see in Chapter 4, is not only the most economical and efficient method of gaining correct information, but the *only* method. The entire course of evolution depends on the incessant appearance of 'mistakes', that is, on chance mutations and the selective retention of those mutations which are more adaptive than others. In biological evolution these mutations are for the most part produced by 'faulty' or imprecise replications of DNA. With the arrival of hominids and especially of *homo sapiens*, many of these mistakes are eliminated or made innocuous by the institution of social cooperation. But the course of evolution continues nevertheless because – though this may sound paradoxical and especially paradoxical in view of the long Platonic tradition that we rely on consciousness for guidance and intelligence – we can depend on consciousness' countenance of three-dimensional language to produce an incessant string of mistakes which are in competition and from which we can select. Consciousness and its linguistic consequence, against all expectation and against common sense, is the great and fruitful source of errors which present themselves for selection.

We can almost reconstruct one of the earliest 'errors' made possible through three-dimensional language in the very early stages of hominid history. Some apes can use sticks as tools. Early man often continued this habit and was able to search for suitable sticks and stones. If at all, two-dimensional language was sufficient for the finding and even manufacture of more or less exact copies of stones which had been used as adzes. But then, suddenly at one point, somebody decided not only to search for a stone which resembled other stones that had been used, but to make one which was different. Failing to imitate, this new shape must be considered an error in reproduction. But some, though most probably not all, of these erroneously constructed adzes proved an enormous advantage. The first human being who made the first 'wrong' adzes – that is, adzes which were not copies of stones which had been found – was the first beneficiary of inchoate consciousness and the three-dimensional language it had pressured him or her and his or her ancestors to develop.

Apart from the ability to generate sentences about objects and events which are not really there and, in doing so, provide a rich repertoire from which to select the least erroneous ones, the ability to generate falsehoods also has an unexpected and tangential spin-off. As was argued above, nature or biology has provided only one firm principle of social bonding: the primary bond and the possibility of its extension in time and over generations by the practice of incest. Societies which are incestuously maintained are destined to remain small and, therefore, economically and militarily vulnerable. To extend societies so that they can include more people also means that one has to have an exclusion principle, so that one can make sure that not every Tom, Dick or Harry can be counted in. A society which is all-inclusive would defeat its very purpose, which is division of labour and defence. The most ready-to-hand principle of non-incestuous social bonding, it was argued above, is false rather than true knowledge. In the past, the vast majority of human societies have been bonded in this way by sets of false beliefs which automatically separate any one society from all other societies which are bonded by other sets of false beliefs. Falsity rather than truth is an effective bonding principle because there is an almost infinite range of possible false beliefs. Two-dimensional language, as we have seen, is not much help in the generation of false beliefs. Even where it enables its users to lie and simulate, we have found that on closer inspection such lying and simulation is nothing much more than a slightly roundabout way of telling the truth. The bird which gives a warning cry about a hawk when there is no hawk is not telling a 'lie' about the hawk, but a truth about the extra food which will be available when the other birds heed the warning and fly away. In order to produce genuine rather than seeming falsehoods, one needs a three-dimensional language. Now the ability to produce false beliefs with the help of three-dimensional language is a direct advantage for social structuring. We must, therefore, conclude that three-dimensional language, in addition to the adaptive advantages it provides in the promotion of knowledge, is also adaptive for non-cognitive reasons.

In this theory of the evolution of three-dimensional language, it is assumed that consciousness precedes three-dimensional, though not necessarily two-dimensional, language. And since three-dimensional language alone is capable of providing a wide range of the systematic falsehoods which are necessary for social bonding beyond pair bonding and incest, we must conclude that the kind of consciousness which drives the evolution of three-dimensional language was selected for *before* the institution of social bonding, which reaches beyond pair bonding and incest. This is at variance with the widely held view that the formation of society precedes the emergence of consciousness and that the evolution of consciousness becomes an adaptive advantage only once society is in place. 'If consciousness is indeed evolved', writes Richard D. Alexander, 'then it must be evolved to enable its bearer to maximize inclusive fitness.'[50] It is widely argued that when, in a society, fellow-members have to be manipulated and related to, conscious beings are at an advantage and that it is the needs of society which provide the reasons for the adaptiveness of consciousness.

There are several replies to and qualifications of this view. To start with, it is certainly true that pair bonding and incest exist without the extra principle of bonding provided by a set of false beliefs – beliefs which, in turn, result from three-dimensional language, which results from inchoate consciousness. It is therefore conceivable that in these very early societies it was already an advantage to have the consciousness which enables members to relate to each other, anticipate each other's responses and manipulate one another. So consciousness could here have been selected for the reasons advanced in this view and indeed be the advantageous result of living in societies. However, one should always bear in mind that inchoate and unarticulated consciousness – the undefined feeling or emotion – is by itself not able to anticipate the responses of fellow-members or make manipulation possible. For this to become possible, consciousness has to be interpreted linguistically as being this or that, and such interpretation is possible only with the help of a three-dimensional language. I would therefore repeat the argument that the primary reason why consciousness was selected for was its ability to exert pressure for the development of a three-dimensional language. Only in tandem with a three-dimensional language is consciousness able to provide the advantages for social behaviour which the view claims are the primary reasons for the fact that it was selected for.

We come, thus, to a seemingly strange conclusion. Consciousness as such, being inchoate and inarticulable as it emerges, has been selected for not because it leads to correct judgements and acts as a guide to emotions and produces goal-directed behaviour, let alone intelligent behaviour. It has been selected and retained because it acts as the generator of three-dimensional language or helps to transform two-dimensional language into three-dimensional language. As it occurs, it is by itself neither here nor there. Like the spandrels of San Marco, it certainly is not an adaptive advantage as it stands. It has turned out to be adaptive because of its consequences. We must, therefore, conclude that consciousness is obliquely, rather than directly, advantageous. It remains to be seen on which conditions the three-dimensional language it generates will lead to more and more true knowledge about the world. In Chapter 1, I will examine the direct consequences of consciousness in its relation to three-dimensional language. The term 'language' will be used when 'three-dimensional language' is meant. It will be argued that this path leads to an impasse because it does not yield a picture of the world as the world appears to itself, but builds up a picture of the world which mirrors consciousness. In Chapter 2, I will examine what happens when we start from the assumption that the causal impact of the world on the organism's nervous system produces in that organism a representation of the world. We will find that on this assumption, the information that enters the organism gets bushed in the nervous system and produces nothing. In Chapter 3, we will start with language and see what happens when we allow its rules to control a picture of the world. We will see that on this path we get a certain distance towards an understanding of how knowledge about the world is generated; but will also find that this path is, in the end, stultifying because it cannot explain the relationship of the vast variety of pictures which can emerge to one another. On this path we get hopelessly bogged down in cognitive relativism. In Chapter 4, I will start with Darwin and examine the consequences of biological evolution and discover that these consequences amount to a vast quantity of knowledge about the world. In the fifth and last chapter, I will attempt to formulate the philosophical adjustments we will have to make, in the light of the philosophical Darwinism of Chapter 4, to our understanding of reality, our grasp of the laws of nature and our understandable ontological anxiety about the universe and our place in it.

1

MAN'S GLASSY ESSENCE

proud man
Dressed in a little brief authority
Most ignorant of what he's most assured -
His glassy essence
(Shakespeare, *Measure for Measure*, II,ii, 117–20)

I

For a long time almost all philosophers have held that if knowledge is a relation between a knower and something that is known, the two sides of this relation are formed by mind and matter. Mental events are events which 'know' the rest of the world, and the main reason for the presence of mental events is the fact that they do the knowing. The world somehow presents itself to the mind and the mind, somehow, represents what has been presented to it. Knowledge of other mental events or minds is a special case. In this model, other minds are either held to be like the knower's mind, in which case knowledge of other minds is a form of self-knowledge, or they are taken to be as different from the knower's mind as non-mental events, in which case the relationship of knowledge is exactly like a relationship between mind and matter. This model has a forceful *prima facie* plausibility. For we do know that there are mental events; and we do know that knowledge is a relation. The origin and persistent plausibility of the model comes from the presumption that these two incontestable facts must be linked together, and this presumption comes from the fact that we constantly talk about our mental events as if we knew what they contained or what they intended or what they were about. This kind of talk creates the illusion that at least some of their 'content' is unalterably known information about non-mental events. If one analyses how we talk about mental events and how our ability to do so comes about, we will see that this presumption is not justified. There are countless variations on this model. But it is no exaggeration to say that from Plato right down to the middle of the twentieth century some such model was used to deal with the phenomenon of knowledge. The model is deeply ingrained. When Richard Rorty in the late seventies became very famous for attacking it, he could not get himself to criticise the model on its own terms and show that, though knowledge is a relation, it is not a relation between mental and non-mental events. If there were knowledge, he accepted, it would be a relation between mental and non-mental events. He tried to show instead that there are no mental events and that, therefore, there can be no knowledge; and then he continued, ambiguously, that when we have knowledge, it merely duplicates what there is and is, therefore, superfluous. First he denied that there is mirroring and then he maintained that there is no point in mirroring. Against Rorty, I will maintain that knowledge is a form of mirroring and, therefore, a relationship and that, though consciousness (or mental events) takes place, it is not one of the terms of this relationship.

We have to revise the model in a different way. It was argued in the Introduction that we know from evolution, first, that there are mental events; and second, that knowledge is a relationship between knower and known and that the emergence of knowledge is anything but a mere duplication of what there is already. In criticising the traditional model we must, therefore, not jeopardise the notion of relationship, and we must take cognisance of mental events. The reason for criticising the model in which knowledge is a relation between mental and non-mental events is a simple one. We know from our studies of pre-human organisms that there is an enormous amount of knowledge in the world and that that knowledge is both acquired and stored without the intervention of mind. Hence, we must conclude that the function of mind, if it is at all crucial, cannot be the precondition of knowledge and that mind cannot be regarded as one term of the relationship. Mental events, when viewed in the light which modern neuroscience can throw on them, are not the sort of events which gather and store correct information about the world. Consciousness is not the sort of condition into which information can flow and by which it can be retained. What, then, are mental events, and how do they function and what role do they play in our knowledge of the world?

Unfortunately the old question 'What is mind?' was off to a bad start. It was off to a bad start when Descartes and Locke in the seventeenth century tried their hand at describing mental events. They gave as examples of mental events such activities as thinking, hoping, believing, doubting, wishing, etc. Attempts to describe mental events and subsequent attempts to relate them to or explain them in terms of bodily events have unfortunately remained bogged down in this initial selection of characteristics. Nobody would question that 'doubting', 'believing', or 'thinking' are mental events. But a closer look will show that these words refer to the way we are doing or holding something, that is, to their modality; not to the fact that the event itself is 'mental'. They are rather descriptive of the modes of mental events. A real mental event is the feeling of sadness or feeling oneself to be a bat or, better, feeling what it is like to be a bat. Thinking about it, believing it to be true, wishing it to be true, doubting it, etc. is not itself a mental event at all, but merely a mode of the mental event. A mental event can be pin-pointed as being 'thoughtful' or as being 'hopeful', and so forth. It can be held to be in the thinking mode or the believing mode, etc. These modes are therefore the adverbial conditions of mental events, not a quality of mental events as such. When 'mental' was taken to refer to nothing more than the mode of doing something, it was easy as well as necessary for Ryle, for eliminative materialists and for identity theoreticians to argue convincingly that to speak of mental events is nothing but a different way of speaking of behavioural events. To label some events 'mental' was rightly and readily considered superfluous – a mere duplication. When we are doubting that the earth is flat, we are not doing two things – doubting (mentally) and behaving by saying 'the earth is flat'. We are really doing only one thing in a doubting way. In modern times it has become fashionable to speak of propositional attitudes rather than of beliefs. But such jargon does not help. It is true that the expression 'attitude' highlights the adverbial quality involved in the stance. But it is unwarranted to identify a mental event with a proposition. A proposition, as I shall argue later, is at best a labelled mental event, not a raw mental event. The distinction between the adverbial mode of behaviour and a genuine mental event is important. If it is not made, it is all too easy to think that since believing, willing, hoping, intending, etc. are merely modes of behaviour or dispositions, there are no mental events at all.

When philosophers mistook the adverbial mode for the 'mental' quality of an event, they predestined their reasoning about mental events to reach the conclusion that mental events reflect or mirror nature and that the cognitive relationship is a relationship between mind and matter. With such reasoning, one always ends up with the conclusion that 'hoping for fine weather' must be a mental event ('hoping') and that that event must refer to something non-mental – 'fine weather'.

There is a corresponding bog which lies at the opposite end of the scale. This is the assumption that mental events occur when there is talk of universals, of meanings, of intentions or of any abstraction whatever. This tradition goes back to Plato but has found a large number of supporters in modern times who believe that mental events are algorithms or patterns or sets of rules for the computation of experiences or the representation of experiences. At present such computer rhetoric is more persuasive than the old-fashioned rhetoric about abstract universals. But on reflection, algorithms are no more mental than universals or the nervous system's power to categorise and recognise similarities in the form of universals, and I would insist that to speak of 'computational consciousness' is a misuse of the word 'consciousness'. The algorithms of a computer are not conscious, and if the human nervous system uses similar or comparable algorithms, there is no reason why we should describe them as 'mental' or 'conscious'. If we do, the words 'mental' and 'conscious' are a pleonasm and do not add anything to the information we get when we refer to these operations as 'algorithms'. Here again one must insist on the distinction between mental events and universals or algorithms. If one does not, one can easily lose sight of mental events, for it is possible to look upon algorithms as dispositions of the nervous system and then conclude that, if mental events are nothing but the algorithms in human or pre-human organisms, human organisms have no mental events because algorithms are not mental.

Hume thought about the matter in terms of a relationship between impressions and ideas, that is, between non-mental and mental events. Mental events (ideas) have their semantic properties by virtue of what they resemble: the idea of John is about John because it looks like him.[1] Brentano's famous characterisation of 'mental' as intentional was an improvement; but only an improvement on Hume's terms. Brentano's intentionality helped us to gain a clearer notion of the difference between mental (ideas) events and non-mental (impressions) events, but this very clarity rammed Hume's distinction even further in. Let us see how it has re-emerged in Searle. He begins his book on the matter with the statement that intentionality is *the* property of mental states.[2] The difference between a non-mental event and a mental event, he seems to be saying, is that the first refers to nothing and the second refers to something. The second represents something. This view is very Brentanoesque, and in so far as it goes a little beyond Hume, it must be wrong. There is a lot of intentionality in the sense of purposiveness in animals,[3] even though it is not verbally articulated and in lower animals such purposive intentionality is not even unworded consciousness: it takes place without any, even pre-linguistic, mental event.

There are many theories about the relation between mind and matter or between mental and physical events. There is the dualism of Eccles and Popper, there is epiphenomenalism, there is the identity theory, there is eliminative materialism and there are several monistic theories. Philosophers bandy their position about with disguised dogmatism. Somebody says: 'I am a monist, and therefore I regard neuronal activity to be both necessary and sufficient to account for mental events.' Somebody else asserts: 'I am a dualist, and therefore I regard neuronal events to be necessary but not sufficient to explain mental events.' Seeing that we are so ignorant as to what constitutes a mental event and equally ignorant as to what constitutes a material event, we can only define the one in relation to the other. But this is like the blind leading the blind, and the contestants must invariably adopt a dogmatic attitude. One could emerge from such dogmatism only if one could define both mental events and material events separately and independently and then study the relation in which they stand to one another. But one cannot do it the other way round, i.e. start with the relation and then assert, dogmatically, that mental events are this and that, and material events, that and this.

By contrast, there are hardly any theories as to the actual emergence or generation of consciousness in human organisms. There are, however, good reasons for that absence. Consciousness by itself is not an adaptive phenomenon. As we have seen in the Introduction, it is only adaptive and has only been selected for because of its by-products. By itself it seems superfluous, and had it not been for its ability to force the generation of language, it would not have been an advantage at all and would, almost certainly, have disappeared soon after it had been thrown up as a mutation. By itself, it has only a negative advantage. It encourages inhibition and, as we shall see later, it helps us to imagine events which have not taken place and even events which cannot take place. In spite of the initial damage it inflicts, especially in regard to our social behaviour and given the ability to imagine what is *not* and to create visions alternative to reality, consciousness has proved a great social advantage. But all this is, if not properly negative, in the nature of damage control and only obliquely adaptive. For these reasons, the conventional strategy of explaining a feature by finding the reasons for which it is adaptive cannot be successful. At best, such explanations in terms of its adaptiveness have to pursue a very roundabout path. Since consciousness is not itself an adaptive feature, it is especially difficult to understand it. 'No one really knows', writes Johnson-Laird in his *Mental Models*,[4] 'what consciousness is, what it does, or what function it serves.' It is not surprising that some of the most famous philosophical efforts of the second half of the twentieth century have been attempts to exorcise consciousness. In the early fifties, Gilbert Ryle called it the ghost in the machine and tried to show that there can be no such ghost because all talk about human behaviour which uses that ghost contravenes logic or semantics or syntax or all three. Correct use of language, he averred, does not allow a ghost to pull strings. The argument was very popular, though I cannot see for the life of me why Ryle's linguistic norms should have been considered correct or why any linguistic norms, correct or false, should be able to pronounce on matters of fact such as whether there are mental events.

More recently the ghost has been seen as a homunculus. Daniel Dennett is famous for an attempt to show how we can do without the homunculus. He exorcises it by pointing out that when we pursue analysis of neuronal computations far enough down, we will not encounter a computing homunculus but only a myriad of very stupid (= mechanical) gadgets functioning one way or another. Talk of a directing homunculus, he is saying, is only short-hand for a vast number of computing processes. The actual events are non-intentional and so stupid that none of them deserves the name 'homunculus'.

II

Let us avoid consideration of dogmatic theories about the relation between mental and physical events and start, instead, by comparing three recent and comprehensive treatments of consciousness: R. W. Sperry's theory of the emergence of consciousness;[5] Daniel Dennett's theory;[6] and G. M. Edelman's theory of consciousness.[7] The theories of Sperry and Dennett are separated by less than a decade, but they present two different theories widely separated by intellectual fashion. Neither Sperry nor Dennett has much concrete evidence to offer, and what makes them remarkably different is, first of all, the rhetoric they employ; and second, the models they use. Sperry wrote just before computer-jargon spread through philosophy, and Dennett is riding high on the crest of the computer-jargon wave. One will gain a good impression of the significance of the rhetoric involved if one notes that where Sperry uses the word 'gnostic', Dennett uses 'cognitive'. 'Gnostic' places Sperry into the old tradition that goes back to post-biblical times, and 'cognitive' settles Dennett right in the middle of the computer revolution. Sperry observes that 'most behavioural scientists today, brain researchers in particular, have little use for consciousness', and Dennett, nine years later, that 'if one looks in the obvious places ... one finds not so much a lack of interest as a deliberate and adroit avoidance of the issue [of consciousness]'.

Sperry called his theory 'emergent interactionism'. In this theory the emergence of consciousness is seen as a holistic event which flows through the entire nervous system. Conscious awareness is interpreted to be a dynamic emergent property of cerebral excitation. The more molar conscious properties are seen to supersede the more elemental physio-chemical forces, just as the properties of the molecule supersede nuclear forces in chemical interactions. Conscious awareness emerges through the hierarchical complexity of the nervous system, and when it does emerge, it affects the entire system so that it makes no sense to think of it as a relation between some parts of the nervous system and other parts. The emergent dynamic properties of certain of these higher specialised cerebral processes are interpreted as consciousness. The cerebral circuits which produce conscious effects may be understandable not in terms of isolated circuit principles, but in terms of advances in cerebral design superimposed on the background of an elaborately evolved central nervous system. Sperry must be aware of the rhetorical element in this description. The rhetoric glosses over the gaps in our understanding, but at the same time conveys a specific message. It states that consciousness emerges at a certain point of complexity and is a condition of the entire nervous system.

Edelman's theory, like Sperry's, is 'holistic', because it concludes that 'the phenomenon of consciousness is the result of a particular order of animate matter that arose relatively recently in evolution'.[8] But it is more detailed and more specific than Sperry's. 'Although consciousness is a process, we shall emphasise', he writes with explicit reference to Sperry, 'that it depends upon the particular organization of certain parts of the brain and not upon the whole brain.'[9] Edelman's theory – the details of which need not concern us here – is able to explain the difference between primary and higher-order consciousness and, therefore, able to show why some animals can be freed 'from the dominance of an immediate driven response',[10] and can conclude that 'through behaviour and particularly through learning, the continual interaction of this kind of memory with present perception results in consciousness'.[11]

Edelman's theory is more ambitious than Sperry's, for it aims to show how we end up with consciousness which has a cognitive content. The basis for this theory is Edelman's attempt to describe the precise nature of the interaction between the world and those neural events which result in consciousness. He argues that the motor activity of the neonate selects groups of neuronal maps for survival inside the body. These maps eventually start to interact, and the evolution of a self is made possible by this selective strengthening of connections within neuronal groups in accordance with the individual's experiences. Edelman repeatedly stresses that the world thus 'experienced' is polymorphous and unlabelled.[12] But if it is, how can it select neuronal groups and lead to the formation of maps? He himself seems equivocal. He says that

> the unlabelled world…is disjunctively sampled by various parallel sensorimotor channels [and that this] sampling results in the selection of combinations of neuronal groups…that are mapped in various ways. Selections of groups within different maps are *correlated* by reentry. For perceptual activity, at least one local map in a reentrant set must receive signals from a given sensory receptor sheet in a fashion which maintains some conformal relation to the spatiotemporal distributions of the real-world things and events that give rise to those signals.[13]

This process of global mapping 'creates a spatiotemporally continuous representation of objects or events'.[14] If the world is polymorphous and unlabelled, it is very difficult to explain how by neural group selection any veridical correspondence between the specific resulting consciousness and the world can be brought about, or, for that matter, how we can give any meaning to the notion of correspondence. The neural system is able to categorise correctly; and it derives this ability from the fact that it has been selected to do so. But in order to establish the mechanism of selection, Edelman slips in again and again the proposition that signals are caused by the spatio-temporal configurations objects and events have in the real world of things over and above what we perceive them to have.

Dennett starts at the opposite end. He gives us a description, following Thomas Nagel, of consciousness as the ability to know what it feels like to be something and then, starting from the top, so to speak, goes down the scale to analyse what kind of functional organisation our neuronal machine would have to have to enable us to utter statements about our own inner self. He assumes, as the touchstone of consciousness, that it is a condition to which we have 'access', that is, a condition about which we can make unequivocal statements. Our verbal productions, he writes,[15] are determined by our semantic intentions. And, in turn, these semantic intentions provide the feelings of the special authority with which we offer our introspective reports of what it feels to be like something.

Sperry, starting from the bottom, works upward, to a picture of consciousness as the effect of the dynamics of neural circuits. The circuits themselves and their constituent sub-parts are not verbally charged and are, themselves, not conscious. Dennett starts at the other end. He begins by telling us what it feels like to be conscious and then builds a picture of the sort of machine which will produce the result he is seeking to explain. He postulates that we have access to our consciousness of feeling what it is like to be something and means by 'access' that we can give a verbal description of it. In this way he is obliged, all the way down, to feel his way from identifiable functioning of neurons to identifiable functioning of neurons. Dennett's consciousness to which we have access is an articulated consciousness. Sperry, starting from the bottom, can afford to think in terms of truly unconscious neuronal circuits which, at the top of a hierarchy, form circuits which make us conscious of what is happening. I would argue that Sperry's model, though couched in out-of-date jargon, is more realistic than Dennett's model, presented in the most up-to-date jargon possible.

Let us look at one specific instance. Dennett[16] provides an example of consciousness: we have a 'presentiment' that someone is looking over our shoulder. We are all familiar with the example; but it is an example which gives a case of *interpreted* consciousness. It is in fact a worded description of an uneasy, inner feeling. Once worded, it ceases to be uneasy and one can even check whether it was justified or not. Now compare Dennett's example to one provided by Sperry. When we are conscious of a colour, Sperry writes,[17] our brain is not adjusting to an array of neural excitations correlated with the colour, but rather to the colour itself. This, he explains, is the difference between being conscious of colour and seeing colour without being conscious of it. The difference between the two examples of consciousness is that to Sperry, the awareness of colour is simply an effect of a highly complex neuronal condition, whereas to Dennett, the awareness of someone looking over our shoulder is the result of an interpretation of a highly complex set of specific computations. Seeing that both Sperry and Dennett are equally ignorant of the specific details, we have to choose between them by comparing the innuendoes of their rhetoric. Dennett's rhetoric is guided by his ultimate hope that one day we will be able to construct a robot which is fully conscious. Sperry's rhetoric is guided by the more old-fashioned concern with gaining a better understanding of what it means to be conscious. It will not help to think that one could restate Sperry's account in the more up-to-date terminology used by Dennett. For the ultimate difference is not one of rhetoric but of the direction of the analysis. Sperry's analysis moves from the bottom up; and Dennett's, from the top down.

Sperry's strength is brought out when one amplifies his description with Jean-Pierre Changeux's characterisation of the process:

> The existence of regulatory loops with reentries at several organizational levels of the brain could lead to high amplitude oscillations... These linkages and relationships, these 'spider's webs', this regulatory system would function *as a whole*. Can one say that consciousness emerges from all this? Yes, if one takes the word 'emerge' literally, as an iceberg emerges from the water. But it is sufficient to say that consciousness *is* the functioning of this regulatory system. Man no longer has a need for the 'Spirit'; it is enough for him to be Neuronal Man.[18]

The two descriptions are compatible and support one another. Changeux's language spells out several neuronal events which Sperry merely refers to. But then, Changeux's rhetoric is in some ways more elliptical than Sperry's, for he compares the emergence of consciousness to the emergence of an iceberg from the water, as if it were not a property of the nervous system but a spiritual excrescence from that system. Both Sperry and Changeux are obviously wending their way along the edges of no man's land, but their charting of the course is helpful. They are using rhetorical expressions because language is inadequate. We will soon reach a point where language will prove so inadequate that not even rhetorical devices will help.

The course, charted somewhat rhetorically, by Sperry and Changeux is helpful in a way in which Dennett's computational rhetoric is not. For it draws our attention to the fact that conscious awareness is a condition of the nervous system and owes its 'emergence' to the enormous complexity of that system. It is a condition of the whole system, and this means that we must not think of consciousness as something that is separate from the occurrences or phenomena we are conscious of. Thus we must not think that consciousness is something which takes place in one part of the nervous system and that that part is conscious of the other parts, which occur separately and constitute the object of which we are conscious. The excitation of the nervous system, regardless of whether it proceeds from outside of our skin or from the inside, emerges as consciousness because of the hierarchically ordered complexity of that system. When we are conscious, there is only one event, not two events as in 'I am conscious of something' and 'there is something I am conscious of'. Moreover, there is only one way of referring to this event, and it is a mistake to think that this event can be referred to either as a material occurrence or, alternatively, as a mental occurrence in the manner in which one can refer to one and the same star as 'the evening star' and as 'the morning star'. By contrast, Dennett's computational rhetoric creates the impression that the verbally expressed consciousness to which we indeed have access is the outcome of analysable, goal-directed functions.

III

With the view that consciousness is not 'consciousness of something…', where the 'something' can be isolated and pointed at separately, we are again facing the inadequacy of language. For in our language the expression 'consciousness' is used to designate 'consciousness *of* something'. The use of 'consciousness' in a non-relational sense is semantically, if not grammatically, wrong. Whether these observations are fruitful for neuroscience remains to be seen; but they are, for the time being, in spite of the fact that they force us into an unconventional use of language, of great help in our philosophical problem. For they make us understand a number of characteristics of mental events.

Dennett's commitment to an established semantic habit is unhelpful. Dennett takes the semantic habit as the ultimate standard. He uses the semantic rule which makes us say: 'we are conscious of something *other* than consciousness' to formulate his argument about the homunculus. In saying that one is conscious of something, Dennett thinks, one is saying that there is a homunculus or a ghost in the machine which is doing the 'being conscious of'. But to know, Dennett continues, that there is a homunculus, one needs a second homunculus who does the knowing of the first; and in order to know that there is a second, one needs a third; and so forth. Dennett believes that this infinite regress proves that psychology is impossible and that one can have no knowledge of one's states of mind. My argument goes in the opposite direction. Since we clearly have, no matter how uncertain, dark and inchoate, knowledge of our states of mind, the semantic habit which leads to Dennett's infinite regress must be wrong or, at least, inadequate.

Let us take a closer look at conscious or mental events. We find that mental events are pure, inward subjectivity. There are philosophers who have tried to consider them the 'immediately given' parts or aspects of knowledge. In a purely temporal sense, they are probably right. But to consider them as something that is 'given' implies that they have a definable and identifiable content and that it is that content rather than the state of consciousness which is 'given'. If we really want to understand the phenomenon of consciousness we must, instead, endeavour to isolate it from that identifiable content. As we are aware of a state of feeling, the very first thing that strikes us is that it feels as if we were at the pit of a dark shaft, the walls of which are utterly smooth so that we cannot climb up and lift ourselves towards the light. Not that having a subjective feeling is necessarily a state of darkness. But it is a feeling of total isolation, associated with the awareness that there are no ladders and no bridges. The awareness of such subjectivity stands at the centre of Kierkegaard's thinking. He considered it an unfathomable mystery. Although the state is awe-inspiring, its occurrence and its specific features do not appear to be all that mysterious. In the *Concluding Unscientific Postscript* he wrote 'that the reflection of inwardness gives to the subjective thinker a double reflection. In thinking, he thinks the universal; but as existing in his thought and assimilating it in his inwardness, he becomes more and more subjectively isolated.'[19] Later he called it 'an objective uncertainty held fast in an appropriation process of the most passionate intensity'.[20] But on other occasions there is evidence that he had to struggle with language to refer to such inward subjectivity, as when he said that consciousness is proof that the self is a relation which relates itself to its own self.

'If we should try to understand that particular self', Proust wrote in his *By Way of Sainte-Beuve,*[21] 'it is by searching our own bosoms and trying to reconstruct it there, that we may arrive at it. Nothing can exempt us from this pilgrimage of the heart ... [into] that world apart, shuttered and sealed against all traffic with the outer world.' Struggling to be more positive, J.-K. Huysmans, thinking of Mallarmé's poetry - 'de mon rêve la nudité' - writes in his *Against Nature*[22]:

> ...taking pleasure, far from society, in the caprices of the mind and the visions of his brain; refining upon thoughts that were already subtle enough, grafting Byzantine niceties on them, perpetuating them in deductions that were barely hinted at and loosely linked by an imperceptible thread.

From these observations about the subjective nature of consciousness we can move to a more modern statement by Thomas Nagel in his well-known essay 'What is it Like to be a Bat?'.[23] To be conscious, he says, is to know 'what it is like to be something'. This statement, if any, underlines the non-relational character of consciousness for any subject, which, clearly, is the sole 'knower' of what it feels like to be that particular subject.

Less sensitive than Kierkegaard, John R. Searle confuses specific subjective states with the occurrence of subjectivity. If science, he writes, is the name of the collection of objective truths we can state about the world, then the existence of subjectivity is an objective scientific fact.[24] The real difficulty glossed over by Searle is that while everybody knows that subjectivity occurs, any one particular state of subjectivity remains inside the person who has it and cannot be communicated, even to the person who is actually having it. This is why Mallarmé called it a state of nudity. In this sense, it is anything but an objective fact.

The heart of the matter, of course, must be introspection. There is no need to consider objections to introspection by people who, like Ryle and Armstrong, do not believe that there are mental events. To them, introspection is not possible because there is nothing to be introspected, and what is described as introspection is in reality only sotto voce behaviour or a form of brain-scanning. But even people who are aware of mental events have become sceptical of introspection, and for a long time now introspection has not only been neglected but also ridiculed. It is surrounded by too many uncertainties, too much subjectivity, too much incorrigibility. Though these doubts are well justified if one expects that introspection will lead to a ready-made package of psychological knowledge, they all come from a misunderstanding of what is involved in 'introspection'. People have expected too much from introspection. They have believed that, by introspecting, one can form an adequate and identifiable account of what it feels like to be something. No matter how much one turns introspection over in one's mind, there is no way in which it can deliver such goods.

However, and this is the heart of the matter, introspection is the only conceivable road to states of consciousness, and it is perfectly reliable provided one realises that these states themselves are not identifiable and specific, so that an identifiable account and a specific definition of such states would actually be a false account and a false definition. The criticisms which have been levelled against introspection as a reliable guide to states of consciousness and the reasons for which it has been rejected are all based on the assumption that introspection, if worth doing, must yield specific and identifiable information. This assumption is both unjustified and unnecessary. The states of consciousness to be introspected are such that no specific account of them is possible, because they are in themselves not specific and identifiable. The shortcomings of introspection, in other words, are its very strength, provided one has a lucid appreciation of the nature of states of consciousness. Before behaviourism and identity theories and eliminative materialism had persuaded people of the folly of introspection, it was believed that introspection could be made reliable by a number of rules.[25] If one avoided, for example, looking at the stimulus that had produced the mental event and refrained from putting an interpretation on the mental event, it was thought that introspection could yield reliable, if somewhat untestable, results. The one area the rules did not consider was the problem of verbal reporting of mental states. And yet, this was the one area which caused the problem. Introspection can lead to absorption in and concentration on a mental state. But it cannot be a guide to the appropriateness of a verbal identification of that state. But since, as we shall see, verbal identification is uncalled for when we attempt to track down a mental state, introspection is a completely proper and dependable guide.

When we are aware, we are obviously not aware of the operations of our neurons. 'No activity of mind is ever conscious', Karl Lashley wrote as far back as 1956.[26] But when we are aware, we are aware of the results or products of the operations of our neurons. These products, however, and this is the crucial point, are totally silent. They may, of course, make the sort of noise caused by vibrations and oscillations. But they are silent in the sense that they carry no labels, do not identify themselves by words, are not articulated in any other way and are not, by themselves, conscious, so that somebody else might identify the state they are and give it a label. These operations are initiated either by activity inside the neuronal system or by an impact of the outside world upon parts of the neuronal system. Whichever way and whatever the complexity of the circuitry which gradually emerges, at no point is that silence ever broken. The products of the dynamics of these circuits, of which we *are* conscious, also remain silent. They come without verbal labels and remain unarticulated until somebody assigns a label to them. Such an assignation, which then defines their specific meaning, is a secondary process and is not the immediate result of the emergence of consciousness.[27] The non-verbal nature of our states of consciousness is their most important and primary characteristic. What we find when we introspect is a deeply inward, ineffable subjectivity which is ineffable not because our language fails us, but because there is nothing specific to be said about it. Its is precisely Kierkegaard's 'objective uncertainty held fast in an appropriation process of the most passionate intensity'.

When we introspect, what exactly do we find? What we exactly find is not exact. As the operations and dynamics of the neuronal system are silent and unworded, the emergence of consciousness is also unworded or non-verbal. As we shall see later, it required a very slapdash assumption to establish the simplistic belief that the silent dynamics of the neuronal system produce states of mind which are amenable to verbal descriptions or, worse still, which magically have a verbal label attached to them, so that each state of mind is semantically identifiable. Without such a slapdash assumption, we find, first of all, that consciousness is self-luminous. This is really nothing but a different way of repeating that when we are conscious, we are not conscious *of* something in the sense that we are doing two things – being conscious *of* a specifiable object. The notion of self-luminosity follows directly from the above description of consciousness as an emergent property of the nervous system. When we introspect, we are simultaneously aware. In this sense, self-luminosity is introspective in substance. But when we try to use language to indicate that substantial event, we are up against the difficulty of grammar and usage. For the word 'introspection', like the word 'consciousness', must be used in a way which indicates what it is we are seeing when we are 'introspecting'. But leaving linguistic usage aside and not treating it as a norm, we can say that introspection leads us to a self-luminous recognition. Introspection, in other words, is its own reward. The activity is all there is.

The consciousness which thus reveals itself is, like the neuronal dynamics which produces it, silent. This means that there are no verbal or any other labels attached to it and that, therefore, its separate states cannot be distinguished from one another and no one state can be identified and specified. Our common usage reflects this condition of awareness by a preference for water metaphors when we are trying to think about it. We speak of 'oceanic feelings', of the 'stream of consciousness' and of 'free-floating' anxiety, love, fear or whatever. Baudelaire, in a famous poem, compared his soul to the waves of the ocean. For water is shapeless and unsegmented, at least in the naive experience of commonsense. In short, the consciousness we encounter when we introspect is inchoate. Being inchoate, it cannot truthfully be reported or described. The feeling one has before falling asleep – when attentiveness is failing and such images one has escape verbal definition and drift, half-formed, aimlessly into one another; and, when verbal definition is attempted in a last, desperate, rallying effort, they vaporise and disperse into the four corners of the universe and fall below the horizon – is probably the nearest 'description' one can get of such inchoate consciousness. It is not only difficult, for obvious reasons, to provide a description. It is also very difficult to actually experience such inchoateness. In our actual waking experience, the verbal labels – regardless of whether they remain purely verbal or proceed towards images – are so close to the inchoateness that we cannot distinguish inchoateness separately. We are so conditioned by education and speech habits that the achievement of inchoateness is really quite difficult and only very rarely spontaneous. It usually requires long training in Yoga or Zen to achieve genuine inchoateness. This, however, should not be taken to mean that it is not a primary state. The reason why it is difficult to achieve is our education. Yoga and Zen are trying to strip that education away rather than build up an achievement of inchoateness.

It has also been noted that our biological organisation requires constant activity and expression, so that human existence in a state of quiescent inwardness is extremely difficult and perhaps not really possible.[28] For this reason, for practical purposes, inchoateness is a hypostatisation we are obliged to make rather than a common experience. In daytime, Yeats observed, our images are unpurged, i.e. locked into strange and ill-fitting conventions which happen to make communication easy. Only at night, when we can take a rest from the exigencies of work and people, can we allow these unpurged images to recede and indulge in the freedom and luxury to purge them of those conventional labels. Nevertheless, even in daytime, we should be mindful of the hypostatisations and of the manner of their presence because of the non-verbal character of neuronal circuitry.

While it has been known for some time, and certainly ever since the age of Romanticism, that subjective and individual experience is ineffable ('*Spricht* die Seele, ach! so spricht die *Seele* nicht mehr!' Schiller wrote; and Wittgenstein, echoing Schiller, remarked less poetically[29] that 'die Meinung fällt aus der Sprache heraus'), the view that these subjective states are essentially, not accidentally, inchoate and are not even articulable by the person who has them, is new. It is, however, not a bizarre view; for it is the only view compatible with the notion that all conscious states are inextricably linked to neuronal events. The inextricability and intimate manner in which they are so linked must ensure that the silence of the neuronal dynamics is transferred to the states of consciousness. If we take seriously the theory that the nervous system is the foundation on which psychic and mental events rest and from which they emerge, we are obliged to assume that the first stirrings of consciousness are inchoate and as such inarticulable. They are so not only to the outside observer, as the Romantics used to suppose, but equally inarticulable to the person who has them. The opposite view, that neuronal events stir up consciousness ready made with verbal labels which we then can report, is quite incredible.

Nevertheless, as we shall see shortly, it is a view widely held by both neurologists and computerising psychologists. The non-verbality of states of consciousness must be a direct consequence of the essential link between neuronal events and consciousness. As we shall see later, in Chapter 4, the impact of the outside world upon the nervous system is quite systematic and organised, because of the post-natal fine tuning of the nervous system. For this reason, the impact of the outside world is not at all inchoate and fuzzy or incapable of differentiation, and quite seriously selective. But however selective and organised, it is unworded and remains unworded right through its transit through the circuits of the neuronal system. If one takes the neuronal background to mentality at all seriously, one is obliged to think of the emergence of articulated conscious events – be they acts of will, of belief, of thought, of dreaming or day-dreaming – as a three-tiered process:

1. Mute neuronal circuitry
2. Stirrings of inchoate consciousness
3. Articulated mental events.

Only by assuming the in-between stage of inchoate mental stirrings (unlabelled emotions or affects) can we explain the transition from the physics and chemistry of neurology to the psychology of a semantically specific mental life. Without such a transition stage, we would remain committed either to an improbable idealism or to a self-contradictory physicalism. Let us try out what the position would be without the transition stage. We would then be committed to explaining our visual perception as follows:

> A series of photons strike the photoreceptor cells in my retina. The signal is then processed through four other layers of the retina and passes through the optic nerve to the lateral geniculate nucleus. From the LGN the signal goes to the striate cortex, zone 17, and then through the rest of the visual cortex, through zones 18 and 19. Eventually this complex electrochemical process causes a concrete conscious visual experience.[30]

Similarly it has been seriously maintained that there is no transition stage because consciousness is 'synonymous' with the mappings and information-processing strategies of information processing.[31] In omitting the transition stage, one gets the sensation of sheer magic as the entering photon comes out, at the other end, as a 'concrete conscious' experience.

With the transition stage we can also understand why the problem of reference of language expressions to the world has proved so intractable. The reference intended or implicit in our speech is a reference to the middle stage, to the inchoate awareness, not to the outside world which makes an impact upon the neuronal system. The relation between the world and the neuronal system is intelligible and organised, and if the world were different from what it is, our nervous system would also have a different shape.[32] But that relation is unworded, no matter how causal. Reference of language, on the other hand, is a relationship between words and unworded, inchoate conscious states. Since they are inchoate, reference cannot be unequivocal and must remain basically inscrutable. Quine is most probably right; albeit for purely neurological reasons, and not because of the philosophical reasons he thinks are so cogent.

Inchoate consciousness may be a novel concept; but the phenomenon itself has been known for a long time. We all know perfectly well what is meant when people speak of a gut-feeling, when Whitehead invited us to reflect on our viscera or when Yeats, more delicately, wrote that 'he that sings a lasting song, thinks on a marrow bone'. Or, as William James put it in *The Varieties of Religious Experience*: 'The recesses of feeling, the darker, blinder strata of character are the only places in the world in which we can catch real fact in the making, and directly perceive how events happen, and how work is actually done.'

The notion of inchoate consciousness is also compatible with our knowledge of animal behaviour and psychology. Konrad Lorenz has stated that in view of the similarities in nervous processes of human beings and higher animals, it would be very strange if they did not have subjective experiences and stirrings of inchoate awareness. The consciousness they are lacking is the secondary consciousness of articulated mental events.[33]

The heightening of neuronal activity makes conscious states emerge, but does not by itself generate labels for it. Since we cannot describe these states of inchoate consciousness, we speak of them as an aura. That aura is undifferentiated, but there are two qualifications to this undifferentiatedness. The first concerns the concept of subjectivity and the second, the notion of a taxonomy in terms of quality.

Introspected, inchoate consciousness is subjective. In one sense this is a truism, because introspection cannot be reported on to other people, except in a very roundabout and slightly mendacious way. It remains, so to speak, where it literally is: inside one's head or, more correctly, inside one's heart or breast. Moreover, it must be subjective, because, as we have seen, there is not really much to be reported, even to oneself. But in another sense, in saying that it is subjective, we are qualifying the inchoateness. For subjectivity is not wholly inchoate. It supposes that there is some sense of an 'I' which is aware, if inchoately so. 'It is futile', J. Z. Young writes, 'to think of oneself as distinct from one's brain.'[34] But how can one form the notion that the brain one is not distinct from is one's own brain? And how can one make an inroad into the inchoateness of consciousness by supposing that there is a self? We suppose that the inchoateness becomes a little qualified by a sense of self because the nervous system experiences its own body and learns to interpret it. A kitten, for example, learns to interpret its visual field by movement. It can do this not by looking at it, but by moving in it. If it had words, it could do it by looking and labelling what it sees. In the absence of words, it moves and associates bumps and obstacles with visual impressions.[35] The neuronally generated consciousness may benefit from similar experiences of the body it belongs to. For this reason we must speak of this inchoate consciousness as a subjective experience.

An inward state of consciousness, though subjective and subjectively known to belong to one's own self, is inchoate. We are clearly aware of something, but we cannot at once say what it is we are aware of. We cannot label the awareness with any degree of certainty or assurance. 'Psychologically, setting aside its expression in words, our thought is simply a vague, shapeless mass ... Nothing is distinct before the introduction of language.'[36] We are aware of a very powerful 'affect', but we cannot articulate the affect, let alone its precise quality. The correct term for

> 'denoting those intrinsic or monadic properties of our sensations discriminated by introspection' is 'Qualia'. The quale of a sensation is typically contrasted with its causal, relational, or functional features ... The quale of a given sensation is at best contingently connected with the causal or functional properties of that state.[37]

The notion that states of mind have a quality rather than an objectivity was already adumbrated by Oswald Külpe (1862–1915) of the famous Würzburg School of psychology in Germany before the First World War. He referred to these peculiar states as *Bewusstseinslagen* – states of the quality of consciousness–and suggested that they are a kind of 'imageless thought', thoughts that occur without any sensory or imaginal content.[38] The quality of itchiness when I feel an itch, the feel of being a bat, the quality of redness in seeing a red rose – all these are introspected awareness of which we are only too strongly aware, even though they cannot be articulated or communicated. The quale is the felt character of a mental state. This indicates that the inchoateness is qualified in the sense that, in spite of the fact that a quale cannot be referred to because of its subjectivity, it has, metaphorically speaking, a certain tone or colour.

We are strongly aware of qualia. But we are not *clearly* aware. And this is where the great catch comes in. In deference to Heller, who has labelled the fact that the world is sane-proof (i.e. not even a sane person can get it right) 'Catch 22', I propose to label the fact that though we are strongly and inevitably aware of qualia, we are not clearly aware of them 'Catch 23'. Catch 23 states that though we find something full of affect and quite powerful when we introspect, we find nothing clear. Our feeling-states, that is, mental events, are initially, and in the raw state in which they are encountered, inchoate; but they have certain colours and tones. Most philosophers and cognitive scientists who have written on qualia have been unaware of Catch

23. They write as if the cause of the quale were known, even though the quale itself cannot be described. Thus it is assumed that when there is a certain quale, we can identify it clearly as 'what it feels like to taste chocolate' when in reality we are not at all sure that the quale in question is caused by the taste of chocolate. They assume, in other words, that the verbal identification of the chocolate tasted sticks to the process that wends its way from the surface of the chocolate via our lips and tongue to the circuits of the neuronal system and comes out at the other end still intact. The chocolate itself, it is admitted, ends up in our stomach, but the quale ends up in our brain and, though no part of the chocolate reaches the brain, the verbal label allegedly does. There is no question that there is a quale. Catch 23 teaches, however, that it is not readily and unequivocally identifiable.

Qualia are much thornier a problem than their advocates realise, and their thorniness does not consist in the fact that there are philosophers who simply deny that they take place.[39] On the contray, it comes from the way they do exist and occur. First of all, one cannot define the taxa to which a quale belongs because of the I/she or he asymmetry of our language. While 'I am sad' has the same grammatical structure as 'she or he is sad', the meanings of the propositions are totally different.[40] The first proposition describes ownership; and the second proposition describes observership. This difference is elegantly obscured by the mere change of pronoun. Second, one cannot form a proposition about a quale the way cognitive scientists have got into the habit of doing. They say that while 'tasting salt' is not a quale, 'what it feels like to taste salt' is. In reality, the quale of feeling the taste of salt or of whatever is so subjective that the second part of the proposition purporting to describe or refer to the quale must remain purely hypothetical and tentative. A quale imagined as this or that, Yeats might have commented, is one of 'the unpurged images of day' which recede at the fall of night.

Though they cannot be identified with any degree of certainty, the existence of qualia qualifies the inchoateness of consciousness in the sense that it enables us to be aware of different colours and tones of consciousness. Thus it is possible to have an initial taxonomy of mental states. The taxonomy is not precise; but it is possible to distinguish between a state which we could, if we verbalised, refer to as 'sadness' from a state which we could refer to as 'joy'. The taxa are broad. It is, for example, not equally easy to distinguish between a quale as 'lust' and as 'love'. Such finer distinctions can only be made with the help of words and images. The possibility of a broad taxonomy is compatible with the finding that any injury to the set of cortical regions located in the dorsal (parietal) region of the cortex results in indifference; and an injury to a set located in the ventral (temporal) regions of the cortex produces a lack of interest in external stimulation.[41]

The possibility of a taxonomy, however crude, is important because it introduces us to the fact that there are extra-linguistic meanings. With mental states inchoate, we cannot expect them to guide any kind of expression, for there is nothing specific enough to do the guiding. But if one can differentiate between mental states, one can speak with Dennett of having 'semantic intentions', for we then 'find ourselves *wanting to say*' something to which these pre-linguistic meanings are a rough guide.[42] The existence of such pre-linguistic meanings is of the greatest importance. Without such guidelines, verbal expression could not be controlled by anything other – and this was Wittgenstein's conclusion – than usage. But given the possibility of pre-linguistic meanings, we can determine, at least broadly, that some expressions are more appropriate than others. We can live up to our semantic intentions. And intention, Wittgenstein has argued, is private and subjective and ineffable and forever behind closed doors. It is not only behind closed doors to outside observers, but also behind closed doors to the person who believes she or he has the intention. For in Wittgenstein's world one cannot go behind language, and any reference to an intention would have to be expressed in language. An attempt to determine whether such a reference is correct or not would lead one to compare one sentence with another sentence. At no time can one compare the sentence with the 'intention'. 'One can say that meaning', Wittgenstein wrote,[43] 'drops out of language; because what a proposition means is told by yet another proposition.' With qualia and their pre-linguistic meanings, the impasse which Wittgenstein believed to exist can be avoided, up to a point and as long as we remind ourselves of Catch 23.

Qualia, even though only the broadest taxonomy is possible, provide extra-linguistic meanings and something which Dennett calls 'semantic intentions'. Dennett, however, it seems, is too sanguine when he believes that these semantic intentions can actually 'determine what we want to say'.[44] Whichever way, we can here detect a bridge between the dichotomously defined possibilities referred to by Hilary Putnam. 'The theory of language understanding and the theory of reference and truth have much less to do with one another', Putnam writes, 'than many philosophers have assumed.'[45] Qualia certainly do not provide a correspondence between words and things and therefore cannot be believed to form a basis for a theory of reference, let alone for a correspondence theory of truth. But in so far as they provide some broadly indicated extra-linguistic meanings, they could be used to learn a language and to teach meanings that are outside language. In extreme cases, they can be thought to provide a touchstone for testing sentences. Suppose somebody who has just lost a fortune on the stock market were asked whether she feels like the person who is dancing with elated joy on the other side of the street. She would almost certainly say 'no' – even though she cannot exactly state how she herself *does* feel. In a very broad, though practically not helpful way, qualia do provide a guide for linguistic expressions.

Like the subjective states themselves, qualia are self-luminous. This expression sounds metaphorical, but ought to be taken in a literal sense. Since qualia cannot become the objects of knowledge – either objects of knowledge for their owner, or objects of knowledge for somebody else – and since they nevertheless appear with a certain 'feel', we must say that they are self-luminous. We know of them directly and non-inferentially when we report our feelings;[46] and such a report, either to the owner himself or herself or to other persons, is the only criterion we have for such consciousness. This has also been put in the following more roundabout way: there is an identity of a conscious state with its non-discursive knowing itself.[47] However, the term 'report', unlike the term 'self-luminous', invites misunderstanding, and here lies another hitch. In saying that the criterion is a 'report' we seem to be implying that such a criterion is a verbal report and that such verbality is an integral part of the criterion. But such an implication is misleading and shows how we are captive to the language we speak – for it is indeed impossible to dissociate the term 'report' from the use of words. Philosophers have made much of the idea that we are captive to wrong use of language – however difficult it may be to state what 'right' use may be. But they have rarely devoted much attention to the fact that we can also be captive to right usage. The usage which makes us say that the criterion of consciousness is a person's report, when 'report' means 'spoken sentence', is certainly right usage. Nevertheless, the idea it conveys is wrong.

This brings us finally to the question of privileged access and of corrigibility. It has long been widely debated whether the conscious states which a person has allow that person privileged access which nobody else can have to them. In view of what has been said above about self-luminosity and subjectivity, the answer must definitely be positive. The person who has conscious states or in whom mental events are taking place has privileged access to them. But when we recall what has been said about inchoateness, it follows that the privileged access is access to no more than an ill-defined quale. So while there is privilege, the access is to something so inward that it is ineffable even to the person who is enjoying the privilege of access. One cannot tell, even to oneself, what one is having access to. Philosophers have always found it hard to accept 'conscious ambiguity' or 'ambiguous consciousness'. They prefer to associate ambiguity with *lack* of consciousness and find it hard to take that privileged access cannot dispel ambiguity. They usually prefer to think that there is nothing there to which we can have private access rather than to accept that we *have* private access, but that it is to an ambiguous or even poliguous state of mind. 'It makes sense to say', Wittgenstein remarked, 'about other people that they doubt whether they are in pain; but not to say it about myself.'[48] Wittgenstein was wrong. I can be very much in doubt as to whether the state of mind I am in is pleasurable or painful, as the entire literature on masochism will amply bear out. The point is very simple: there is a state of mind which we come across, so to speak, when we introspect and of which we therefore say that it is self-luminous. But that state comes without a verbal label and, though it is self-luminous and a state to which we have private access, we may well doubt whether any verbal identification we attach to it is correct. States of mind, once verbalised, are very corrigible. The difficulty is that they cannot be corrected once and for all. They are corrigible; but there is no clearly defined state we can compare them with so that we may conclude that they correspond to this state or not. They have an aura or quale; but the quale is unidentifiable and any corrigibility remains suspended in doubt, even though it makes very good sense to say that they can be corrected. For in reporting them, we may well be wrong and identify a state as painful when it really is closer to pleasure.

A statement about a sensation, like having a pain, Wittgenstein said, cannot be corrected. There cannot be a sensation over and above 'having a pain' and one cannot, therefore, say that the statement is false – unless one is consciously lying. But Wittgenstein telescoped the problem. The sensation in question is a quale. That quale cannot be corrigible, because apart from being the way it is, there is nothing one can compare it to in order to be able to say whether one has that quale or not; except if one mendaciously claims to be having it when one does not. But the quale comes without words. When one is saying 'I have a pain', one is saying more than that one is having a quale. One is attributing a verbal label to it. Now, that attributed label can be the right label, or it can be a wrong label. The situation is difficult to decide, because the qualia are only very broadly, but not specifically, classifiable. Nevertheless, it does make sense to ask whether the attribution is correct or not, and that makes the statement corrigible. The corrigibility, however, only concerns the verbal label. A wrongly attributed label can be corrected when one is using the pre-linguistic or extra-linguistic meaning enshrined in the quale as a touchstone.

An individual, subjective consciousness is so inward that, in spite of the presence of qualia, which tend to give it a tone or a colouring, the inner subjectivity is an airy nothing without a local habitation and without a name. This makes it ineffable, and such ineffability applies to the person who is aware of that inward subjectivity as much as to any other person. Such consciousness may be a comfort, but a cold comfort at most. It is as if we are aware that we are boxed in. We are uneasily aware that we are and what we are. But we cannot be sure what it is we are, and we cannot say so, and we cannot even whisper it to ourselves.

Nevertheless, a quale has an aura or a shade of meaning, and, though we cannot present a taxonomy of qualia, we would not be justified in calling them qualia if we did not mean to say that they have vague qualities of meaning. Wittgenstein is famous for having remarked frequently that there are no extra-linguistic or pre-linguistic meanings in terms of which we can check whether language is being used correctly or not. Sellars has called the view that awareness comes first and that language follows and that the use of language can be tested as to whether its use is or is not adequate to the awareness, the Myth of the Given. Rorty has commented that 'adequate to' is an empty notion because one cannot compare a bit of language to a bit of non-linguistic awareness.[49] But the colours of qualia will out, and, though they are not verbally identifiable, will strain to express themselves in a private language. According to Wittgenstein, there can be no private language because to use language is to follow a rule; and, clearly, there can be no private or idiosyncratic 'rules'. But there is plenty of evidence that private languages have been used and used in plain defiance of rules and that the special extra-linguistic or pre-linguistic meaning of qualia has been expressed by the fact that rules are *not* followed. Or, since the languages we speak are bound by rules, should we say that the privacy of the expression is conveyed by the idiosyncratic manner in which the known rules are broken?

Wittgenstein, we will recall, stated that 'meaning falls outside language'. But nevertheless, he sometimes tried hard to use weird language to convey precisely that kind of meaning. He wrote, for example, that when, on seeing a rabbit dash past, one exclaims 'a rabbit!', one is not just proving that one has learnt to use the word 'rabbit', but that one is 'thinking' that one should use the word'rabbit'. On exclaiming 'rabbit' when one sees the flash of a darting rabbit, one is combining thinking and seeing, that is, one is combining two private, mental and subjective events: half experience, half thought.[50] ' "The echo of a thought in sight" – one would like to say'.[51] Wittgenstein was by no means the first to try his hand at private language, though he was the first philosopher to do so while proclaiming that it cannot be done. Age-old metaphors – e.g., the sight of sound – are attempts to convey pre-linguistic meanings by going against the rules of language, and poetry often consists of nothing but attempts to sponsor such metaphors. The most sustained and systematic effort at private language was made by James Joyce. At first in *Ulysses* he structured it by following the direct stream of consciousness and thus purged it, even in daytime, as Yeats would have said, of those rational images and their logical sequentiality we invest them with and tie them up in. And, later, in *Finnegans Wake*, he not only purged language of its rationality but constructed his own words by allowing free association to elaborate or distort their meanings[52] to knot the introspected, self-luminous qualia together 'with an adhesive style, a unique hermetic language, full of contracted phrases, elliptical constructions, audacious tropes'.[53]

IV

Before we go into the consequences of these findings, we must look at the enormity which has resulted from their neglect. This neglect has found expression in the slapdash assumption that human neuronal systems have semantic properties. Consciousness, it has been widely agreed, is a very mysterious phenomenon. Its mystery comes largely from the fact that in so far as it is pre-linguistic and unworded it also tends to appear as something separate or only loosely linked to neuronal events, and this appearance appears as an aura which is non-material and spiritual or pneumatic, as the ancients put it. But the spiritual appearance is deceptive and comes from the absence of words, not from absence of neurons. It is phenomenal, not ontological. As if this deceptive appearance were not enough, conscious states become doubly mysterious if one takes it for granted that in all cases they are inextricably wound up with a verbal report or linguistically identified. Many philosophers, in fact, go so far as to think that we 'mean' by consciousness a verbally expressed and identified state of mind. Dennett's slapdash identification of 'consciousness' with 'what we have access to'[54] is by no means an isolated example. He is not only in good company, but also in the company of just about everybody. It was part of Hume's tacit assumption that sense-impressions are magically transformed into words, and he might, had he been pressed, have believed that all sense-impressions are 'worded' or tagged with words – so that when we see an orange, we can have a sense-impression of the orange and then transform that sense-impression into the word 'orange'. The assumption was tacit because in Hume's day nothing was known about neurology and neuroscience did not exist. Hence, had he been questioned, Hume would not have thought that any kind of magic is needed to produce the wording of sense-impressions. We must certainly forgive Hume. Not only is neuroscience very much a twentieth-century achievement, but even avowed practitioners of neuroscience, as well as neuroscience-oriented philosophers who now call themselves 'neurophilosophers', have not been able to appreciate the size of the problem. And for that matter the initial Humean belief that sense-impressions magically transform themselves without further ado into words was still shared in the early decades of our own century by the Vienna Circle, who believed in 'protocol-statements' – that is, in the existence of observations that were, provided the observations were simple enough, automatically transformed into verbal expressions.

John R. Searle's description of the ascription of verbal identity to a state of mind sounds more subtle. 'The mind imposes', he writes,[55] 'Intentionality on entities that are not intrinsically Intentional by intentionally conferring the conditions of satisfaction of the expressed psychological state upon the external physical entity.' It is not clear whether Searle means by 'physical entity' the outside event or object or the inner neuronally induced state of mind. But in any case, the use of the term 'satisfaction' makes the process obscure rather than subtle. When does satisfaction take place? And if satisfaction is dependent upon an 'intentional conferment', where does this second-order intentionality come from? And if satisfaction takes place when there arises a correspondence between an expressed psychological state and an external physical entity, are we to think of this satisfaction as coming upon us magically, or is it produced by a logical check on whether there actually *is* a correspondence? And if the latter, how can we know what the expressed psychological state *is* before it has been stamped intentionally upon an external physical entity?

None of these opinions or tergiversations can work. The world gives signals or stimuli.

> While the world is not amorphous and the *properties* of objects are describable in terms of chemistry and physics, it is clear that at the macroscopic level, objects do not come in predefined categories, are variable in time, occur as novelties, and are responded to in terms of relative adaptive value to the organism rather than of veridical descriptions. This lends a relativistic and disjunctive flavor to the categorisation of objects by animals.[56]

The signals picked up by the nervous system are silent and unworded and remain so inside the body, no matter how complex the circuitry and the re-entrant remapping is. It cannot go without saying that somewhere along these re-entrant mappings, words attach themselves to these neural events. When these silent states of mind, no matter how self-luminous they are, become identified by words, we presume again that neuronal events must be involved. But the circuits involved in this process are very different from the ones which brought about the state of mind. For in the wording procedure there is manifest and explicit semantic intention or content, and all specifying meanings come ready-made. Language use, unlike neuronal events brought about by outside signals, is *not* silent. A language and its catalogue of ready meanings are learnt by being part of a speech community. And such learning is a very different matter from picking up amorphous signals from the outside world and putting them through their neural paces until a state of inchoate consciousness is generated. The real problem and the sixty-four thousand dollar question is how language and these states of mind can be brought together. Nothing can be gained by confusing the issue sophistically or inadvertently by saying that they belong together either automatically or magically. And above all, we should guard against what are sometimes described as 'biofeedback techniques' and alleged to have created a science of subjective experience.[57] It is one thing to *be* self-luminously aware of a state of mind and quite another to detect alpha waves or rapid eye-movement which accompany it.

To be sure, Newton knew that there is a problem, and often called red light 'rubrifick' (i.e. 'red-making') because it consists of nothing but rays which are disposed to propagate a movement in the sensorium in which they are sensations of those movements under the form of colour. Even though both Locke and Newton knew that one must make a distinction between wave-lengths that impact upon the body and colours we are conscious of, they were not aware that there ought to be a second distinction – the distinction between the subjective awareness of colour and the attribution of a worded label to that awareness. We are most probably conditioned to identify a certain awareness by a certain word; but the rules governing the employment of the word are very different from the laws which govern the emergence of the subjective awareness. With the help of a word, we drag the subjective awareness into the public arena. But the coming together of a word and an awareness is like an alliance between two independent rule systems and cannot be taken for granted. Our knowledge of when to deploy the word 'red' comes from the language-game we participate in, whereas the awareness we label by that word comes from the lengths of the light-waves we sense.

The belief that the second distinction is unnecessary and that, given the sensation, the word must emerge by magic is indeed deep-seated and will not die. It must be one of the tacit assumptions behind the so-called Received View of the philosophy of science according to which scientific theories are axiomatic calculi which are given a partial observational interpretation by means of correspondence rules.[58] There is nothing in any of these correspondence rules to govern the transformation of physical or chemical stimulation of the nervous system into articulated verbal expressions. That transformation is not mentioned and seems believed to be magical. If the Received View is innocent of neuroscience, the same cannot be said for Patricia Churchland. In her book entitled *Neurophilosophy*[59] she states that conscious and, presumably, verbal expressions of colour, taste and odour have as their 'physical substrate a neuronal phase space whose axes are [in the case of colour] three streams of neurons carrying information about reflectance efficiency values at three wavelengths'. Based on Land's results,[60] her statement is on firm neurological and optical ground. What is amazing and daring is the jump into what is after all a worded awareness and a verbal expression. She makes this jump with tacit ease as if it went without saying that once a neuronal pathway is discovered, the verbalisation follows automatically.

Or let us look at the way a well-known neurologist treats the matter. Jean-Pierre Changeux says, for example, that 'mental objects with a realistic component, like pictures ... mobilise neurons in the right hemisphere, while those with a more verbal content – concepts – recruit neurons in the left hemisphere'.[61] Changeux maintains that this is evidence that the barriers that separate the neural from the mental have been destroyed.[62] He provides other examples to show what he means by 'destruction'. There are, he writes,[63] areas where cell bodies and processes contain a specific neurotransmitter – *dopamine.* 'In the hypothalamus and the brainstem, dopamine synapses can thus be regarded as "pleasure" or "hedonic" synapses where the cold information regarding the physical dimensions of a stimulus is translated into the warm experience of pleasure.'[64]

Changeux's view is seductive; but his rhetorical devices disguise the real difficulty. He says that neurons are 'recruited' and that 'cold information' is translated into 'warm experience'. There is no justification for these rhetorical leaps. Dopamine, undoubtedly, generates a certain feeling inside the body. But that feeling is opaque, inchoate and blind, because unlabelled. There is no reason why we should simply assume that the term 'pleasure' is the *correct* label. Indeed, as is well known, many of those feelings would respond equally well to the term 'pain' – hence, for example, the psychological phenomenon known as 'sado-masochism'. Whatever mental awareness is neuronally generated is an unworded or unlabelled awareness. The identification of such awareness by the ascription of a word such as 'pain' or 'pleasure' cannot go without saying, for neurons do not carry labels and their circuits do not flash verbal messages. Even though neuroscience has made progress in bridging the gap between neuronal and mental events, the belief that it can also bridge the gap between physical events such as neuronal circuits and mental representations is therefore illusory. There is no easy bridge from neuronal substrates to verbally labelled mental images and concepts.

The euphoria of some contemporary neuroscience was due to important discoveries. It was one thing to discover that there are certain areas of the brain concerned with vision and that injury to a set of cortical regions, located in the dorsal (parietal) region of the cortex, was critical for surveillance, attention and arousal: its injury results in indifference and in the '*loss of a sense of caring about one's own person.*'[65] But it is quite another to conclude from such evidence that there is a one-to-one correspondence between neuronal events and the verbal designation of the inchoate feelings they produce. As stated, we are confronted with three sets of events. There are, first, the neuronal circuits; second, the inchoate or blind feeling (emotion?) they generate; and, third, the verbal identification of that opaque feeling as 'orange' or 'pain'. The first two events are, as neurological evidence now indicates, very intimately linked. But the third type of event is a hypothetical superimposition and cannot stand in a causal relationship to the two other types.

Consider, for example, the very striking contribution made by David Marr to the theory of vision. Marr's investigations have shown that there is direct energy transfer from the object seen to the nervous system. He discovered the emergence in the nervous system of the 'primal sketch', which is a primitive but rich description of the intensity changes that are present in an image.

> All subsequent analysis reads the primal sketch, not the data from which it is computed. The primal sketch therefore acts in a genuine sense as the interface at which visual analysis becomes a purely symbolic affair ... Downward-flowing information will not affect the line-finding stage ... at all. Its most usual *modus operandi* is in choosing which processes are to be used to read the primal sketch ... [66] [There is] acquired knowledge which is stored and catalogued in our brains ... The problem of searching for catalogued information ... occurs only

> at this final stage of visual processing. Nobody had imagined that so much information about shapes could be extracted from the retinal images before a search of catalogued information would be necessary.[67]

Marr's work documents in great detail the relationship between input and final awareness and the way the nervous system interacts with and computes the input and demonstrates beyond all shadow of reasonable doubt that the input is no more than a primal sketch of the object seen and that what we call vision is the brain's computation and analysis of the data provided not by the object seen but by the primal sketch it causes. But not even this detailed account of the process of vision can bridge the last gap between neuronal computations performed on the primal sketch and the moment at which we say, e.g., 'chair', that is, the moment where we provide a word. Marr insists that the primal sketch corresponds very closely with the image we are conscious of, i.e. with the moment at which we say 'chair'. But close correspondence is not the same as identity and really can mean no more than that the primal sketch does not falsify the moment at which we say 'chair'. Somewhere during the process of downward flow of information to the primal sketch there occurs a verbal superimposition which is not part of the neuronal computation. Or, in so far as it is part of it, it uses information stored in the brain from a source other than the primal sketch.

It seems indeed that the most important contribution neuroscience has made so far to our old philosophical problem is that it has explained how energy is transferred from objects seen or smelt or touched to the nervous system and how this energy is processed inside that system, so that we can be confident that perception of something real does take place and that the events in our nervous system are related to something *real*. But at the same time neurological inquiry into nervous circuitry has also demonstrated that at no stage in this circuitry are the events demonstrably verbal nor do they become so. In all cases, they remain neuronal. We can get from outside objects to nervous events and from nervous events to feelings and awarenesses. But these feelings and awarenesses are and remain unworded. We are thus left, from neuroscience, with the assurance that the outside world makes a very direct and traceable impact upon our nervous system; that that impact is noticeable in terms of opaque and inchoate feelings; but that these feelings are indeed opaque and inchoate, because non-verbal. This is a step forward; but in a direction which can bring no comfort to Hume and his followers. There is every reason to suppose that these neuronal circuits mobilise or activate neuronal circuits in which the appropriate words are encoded until they are uttered. But the point is that the source of the verbalising circuits and the source of the circuits which compute Marr's primal sketch do not coincide. The first circuits come from a language-game we participate in; and the second, from the image or object in front of our eyes.

Once this is accepted, it follows that not even contemporary neuroscience can solve Hume's problem. Long before we reach the heights of theoretical science, we are confronted by the fact that there is no straight line which leads from impression to verbal expression. Even quite ordinary experience is fatally divorced from the verbal description of that experience, even though nowadays, with the assistance of neurology, we can understand how and why objects outside our body worm their way into our nervous system and generate these feelings and opaque emotions. These feelings, however, are as unlabelled as the neuronal circuits which produce them. All we can say – and this is probably an advance over Hume as well as over the Vienna Circle – is that the gap which we cannot bridge does not lie between the outer world and our body-mind, but between the circuitry of our nervous system and our language. Neuroscience has shifted the problem from one place to another, but has not solved it. In that sense, we are still with the difficulty which Hume had not been able to cope with.

The progress of neuroscience has been to track down pathways of neuronal functioning and even to establish, thinking of the great work of Edelman, how these pathways are naturally selected in response to the external world. But, at the same time, neuroscience shows that these pathways are always silent, not only in Lashley's sense, but also in the sense that they have no semantic properties. Neuroscience is therefore cold comfort to eliminative materialists and identity theories on one side and to spiritualists, dualists and epiphenomenalists on the other. The upholders of such views all believe mistakenly that neuroscience demonstrates either that causality does not extend beyond the system of neurons and does not need to, because the neurons somehow have semantic properties; or that causality somehow must extend beyond neurons because the latter have *no* semantic properties. It seems that the more correct conclusion from the findings of neuroscience is that since neurons have no semantic properties and since we do talk and think in words and images, the connections which extend beyond neurons are *hypothetical interpretations* of neuronal events. In this case, the question of any interaction between the mentally defined world of symbols, be they images or words, and the world of neurons does not arise. The world of hypothetical interpretations is not causally linked to neuronal events and does not causally interact with them. In rejecting the slapdash assumption, we are left with three tiers:

1. There are neuronal events, caused by other neuronal events or by impacts from the outside.
2. There is an inchoate and silent aura of consciousness, caused by neuronal events.
3. There are hypothetical definitions of the inchoate aura.

The verbal definitions of (1) also could be taken to relate to and include the causes of (1). Hence the verbal expressions of (3) actually constitute and define the events in the outside world which caused the events of (1). But this relationship is very different from what people mean when they speak of reference. One can only speak of the 'reference' of A to B when one has independent knowledge of B, so that one can determine whether A *really* refers to B or to something else. In the case outlined, however, there can be no such independent knowledge of B, and the question whether there is reference of A to B cannot make sense: for B is constituted by the hypothetical definition of the neuronally caused aura and is not identifiable separately and without it. Hypothetical, verbal statements are about states of inchoate consciousness; not about events in the outside world. The causal chains set in operation by the world come to an end inside the neuronal system. The interface between that world and the hypothetical definitions is between (2) and (3) and not, as the slapdash assumption would have us believe, between (1) and (2). If we reject the slapdash assumption, how can we explain that we break out of the box all the time?

V

The explanation is that we assign verbal labels to the inchoate states of consciousness. Let us begin with an example from Wittgenstein. Wittgenstein noted that we may see odd lines and then, taking a second look, we may see them as a familiar face.[68] If we were to make a drawing of the first perception and then of the second perception, he goes on, the two drawings would be identical. This indicates, he concludes, that though the perception (i.e. the state of mind neuronally set in place by whatever there is in the world) does not alter, the description or identification of what I see does.[69] The importance of labels is that they make us *see* the things we label as if the label had not intervened. However, we cannot compare what we *see* with the label we assign to it. If we did try a comparison, we would have to make a drawing of the unlabelled perception. But if we did, as Wittgenstein ingeniously pointed out, we would always make the same drawing, regardless of the label we use to interpret it.[70] As is well known, Wittgenstein concluded from this and similar observations that there can be no private language or pre-linguistic meanings and that the meanings of our sentences depend on the rules governing their use, not on their correspondence to perceptions. Since the states of mind in question (caused by the lines in front of us) are, among other things, qualia with pre-linguistic, if broad, meanings, none of these conclusions really follow. But the initial observation is of the greatest importance.

The labels we most commonly use are words and sentences. The Romantics believed that words are too inexact to express or refer states of mind. My argument is that it is the other way round. While the meanings of words are well known and their usage quite exact, it is the states of mind which are too inexact to be fastened unequivocally and unambiguously to words. States of mind are – and this derives directly from their inchoateness – malleable to identifying verbal articulation. Their precise meaning, as distinct from the possibility of a broadly hinted taxonomy of qualia, depends on vocabulation, or better, depends in the first instance on vocabulation. Take any state of mind like 'loving'. It is well known that one often cannot say whether this should be identified as 'love' or as 'hate'. When people say: 'I love my country', it is well known that, on analysis, it can mean, in terms of their feeling-states, 'I hate my father.' If we have a neuronally induced vision of a tree, it is often quite difficult to know whether this is to be seen as a tree, or whether it should be a cross or possibly an eagle. This means that the malleability of states of mind and the need for vocabulation is not confined to those states of mind which we class as emotion, but applies equally to those which we are accustomed to call 'perceptions'.

'The physical nature of the neurophysiological substrate', writes Johnson-Laird, 'places no constraints on the patterns of thoughts.'[71] This means that the labels we assign to states of mind are not caused by what goes on inside any more than they are caused by what goes on outside. Indeed, the causal connection between what goes on outside the body and what goes on inside the body is quite close. The physical substrate is caused. But the vocabulation and symbolisation are not. The sight of an attractive woman causes a substratum of libidinal energy 'in his mind'. But it requires a lot of culturally conditioned articulation and labelling before this malleable libido becomes the phenomenon of erotic desire or romantic infatuation. The causal relation between events in the outside world and the malleable states of mind is quite stringent. The malleability of the latter derives from the unworded nature of the neuronal events, not from any lack of precision with which the signals are picked up from the outside. On the contrary. We know now, thanks to G. M. Edelman, that the nervous system, through post-natal fine tuning through selection, is quite well adjusted to the structure of the messages it comes across, for we have learnt to classify events by a 'selective system in which a large pre-existing set of variants of neural networks formed during embryonic life ... are later selected for and against during the worldly life of the animal'.[72] None of this implies that the precision of the picked-up message includes a verbal label.

The label assigned to the mental event in vocabulation is chosen with a great deal of freedom because not even the taxonomic identity the mental event possesses in virtue of its character as a quale is sufficiently strong to *force* us to adopt one particular label as against another. We are familiar with the uncertainty of introspected states of mind. Am I *really* in love? Or do I really love *this* woman? Or do I really *love* this woman? In conventional society one asks: what are your intentions? But equally to the point, ought the more unconventional question be asked: what are *my* intentions? The assignation of a verbal label is therefore fairly much a free choice and, given the malleability of the mental event, that free choice must always remain hypothetical. Since we cannot identify the mental event by itself as we inwardly experience it, we can never be sure which label is most suitable. Hence, the reference involved in the choice of a label remains inscrutable, and St Thomas Aquinas' realisation that for a state of consciousness to become defined there has to be a determined and intentional *conversio ad phantasmata*[73] is very much to the point. As was argued above, Quine's inscrutability of reference does not concern the relation between mental event and outside occurrence; but the relation between the verbal label and the mental event. By contrast, following Edelman and Changeux, the relationship between outside occurrence and mental event is causal and fairly specific because the nervous system has evolved, both before and after birth, in response to that outside world and to the encounters the organism is likely to have with it.

Jerome Bruner has called the culturally generated and transmitted component – as distinct from the physical input of the outside world – of our motoric and reflective capacities the 'amplifiers'.[74] One can apply the concept 'amplifier' equally well to the linguistic label we attach to the neuronally generated states of mind. However, and this is an important point to which I shall return in Chapter 3, these particular amplifiers, though also supplied from the outside, are not supplied in tandem with the physical input of the outside world which generates the inchoate states of mind. Instead, they come from behind, so to speak, as part of the social culture the organism finds itself in. The reason why we should regard them as amplifiers is that they define and specify inchoateness. Also, these labels supplied from behind, by culture, come in sets. They form part of a belief system and a set of idioms which reflect those belief systems. When one assigns one label to one state of mind, one commits oneself, therefore, automatically to an entire set of labels which define other inchoate states of mind. In certain cultures, for example, if one commits oneself to defining a certain state of mind as 'loved by mother', one also commits oneself to defining another state of mind as 'ontic security'. While the culture from which the labels are drawn imposes constraints on the labels one may choose, the essential malleability of states of mind remains. The very origin of culture and the rules it imposes is linked to the ability to resolve the thrashing around for labels which must occur because of this malleability.

This view of labels amplifying inchoate states of mind again underlines the problem of the correctness of such labels. The label gives definition and meaning to the state of mind. Being malleable, that imposed meaning cannot be the right meaning any more than it can be the wrong meaning. Better, it cannot be the right meaning; but it can, given the fact that the state of mind is a quale with a rudimentary pre-linguistic meaning, be the wrong meaning. While it is impossible to be certain, for example, as to the precise label to be used for the distress of the woman who lost her fortune on the stock market, one can be more certain that a label which states 'satisfaction' would be the wrong label. One cannot, however, go beyond this hypothetical nature of the assignation of labels. If one wanted to know what would be the correct meaning, one would have to be able to form a clear picture of the inchoate state of mind *before* it is labelled. To do this would involve labelling the state of mind. One could then compare the state of mind with the label and decide whether the chosen label is the correct label. Given the inchoateness of the state of mind, this strategy is not available. This situation must, therefore, lead to the view that labelling is comparatively relative. However, it is only the labelling that is relative. The inchoate state of mind is not. It does stand in an intelligible causal relationship to the outside world and/or the inner condition of the organism. Its ineffability links it to the procedure of labelling, which is a relative procedure – relative, that is, in relation to the cultural rules the label is taken from. We find here a coming together of a causally determined process and a procedure which is relative to a given culture. As far as our knowledge of the world is concerned, there must be a way to resolve this apparent conflict between causal determinism and the relativistic interpretation of the products of that determinism. We shall face this problem in Chapter 4.

While we can have no difficulty in recognising inchoate states of mind as mental events, many people have experienced great difficulty in recognising consciousness of a chair as a mental event. To make the labelled state of mind appear as a mental event, one would have to be able to say that one is conscious of being conscious of seeing a chair. There is clearly an infinite regress here. To avoid such an infinite regress, it is often alleged that 'being conscious' is not a mental event at all, but merely one of many possible modes of seeing a chair. This reasoning does not take into account the distinction between the inchoate state of mind and the labelled state of mind. Inward, inchoate consciousness is final. It is inconceivable that there should be an infinite regress. One cannot be inchoately conscious of inchoate consciousness. Being inchoately conscious of inchoate consciousness is in no way different from being inchoately conscious. However, as soon as the initial state of mind is labelled, the infinite regress becomes possible. One can very well be conscious of being conscious of seeing a chair, and so forth. The infinite regress, though, does not make nonsense of the expression 'I am conscious of a chair.' What it does is something quite different. It brackets the chair, and then it brackets the bracketed chair, i.e. 'seeing chair', and so forth, until the original label 'seeing a chair' is so deeply bracketed that it is almost lost sight of. It becomes so encapsulated that we remain, after performing a reasonable amount of bracketing, with what is for practical purposes an unlabelled event again, once more a very nearly pure condition of unlabelled consciousness. This shows that the infinite regress and its outcome concern the *label*, not the mental event in question.

Stephen Stich, in his excitingly observed but wrong-headedly concluded *From Folk-Psychology to Cognitive Science,*[75] is aware that there is something airy about having, say, beliefs. He senses that, somewhere, the presence of beliefs implies that there is some uncertainty, some lack of one-to-one correspondence. But he locates this airy uncertainty in the wrong place. He says we *hypothesise* that belief is a relation between a person and an internally represented sentence.[76] He ought to have concluded that it is the relation between the person and the internally represented sentence which must remain hypothetical, because what he calls an 'internally represented sentence' is in fact, as neuroscience obliges us to think, not a sentence at all but a merely silent feel of an unidentifiable pure quality. Instead he jumps to the conclusion that it is *we* who 'hypothesise' that belief is a relation between a person and an internally represented sentence. The belief that a belief is a relation between a person and something internal (one ought not to think of that something as a 'sentence'!) is not in the least 'hypothetical'. This belief, on the contrary, follows from 'having a belief'. The hypotheticality enters into the situation at the point at which a worded label is assigned to the internal state which is thus transformed from a mere quale into a hypothetically worded statement.

The view presented has an interesting implication which has been given much prominence in contemporary literary criticism. If labels are hypothetical attributions, the meaning of a state of mind cannot be determined by authorial or owner's intention beyond the rudimentary meaning it has in virtue of being a quale. For the author or owner of the state of mind does not know what it is before the label is hypothetically attributed. In that case, the author or owner of the state of mind has as much or as little right to make an attribution as anybody else. The modern school of deconstruction has made the most of this situation by claiming that verbal signs have a life of their own apart from their authors' intentions. Described in this way, they seem to be brushing aside an author's legitimate proprietary right in the meaning of what she or he is saying. In reality, however, that right need not be brushed aside, because it does not exist. The author herself or himself does not know what she or he is intending to say until she or he has attributed a label to her or his intention. The label is hypothetical, and the attribution of labels is open to anybody. A text is therefore *essentially* indeterminate and open to interpretation, and not just indeterminate because the authorial privilege of interpreting it has been wrested from its rightful owner. Wittgenstein proved more perceptive here than Derrida. Wittgenstein recognised that authorial intention cannot be determined, because such a determination would necessitate a recourse to something other than language. Derrida merely argues that the authorial intention does not constitute a privilege. For Wittgenstein, it cannot constitute a privilege because it does not exist.

The first or any label assigned to a state of mind is not a literal description, even a tentative one. A true literal description of the state of mind would have to be a true description of an ontological subjectivity. 'Truth' must involve correspondence. One cannot make a linguistic expression correspond to a non-linguistic event. Even if one were to apply the Tarski formula and insist that 'I am sad' is true if and only if I am sad, it would not make sense. For in order to apply that formula there would have to be a specified procedure for finding out whether I am sad or not. But since the state of mind referred to is unworded and, in its unworded condition, incorrigible, nobody can know whether it is correctly described as 'being sad'. Even leaving this question aside, there is simple proof that description, being bound up with language and its rules, canot do justice to the phenomenon. In our language we can only describe the state of mind by saying 'I am sad' or, if somebody else is watching me, 'he or she is sad'. The structure of these sentences is equivalent. One simply replaces one pronoun for another and thus is making it look as if the statements had comparable meanings. In fact, however, their meanings ought not to be comparable. When I am sad, I am identifying my own mental state. But the unworded state itself is very passionately manifest inside me as my own subjective consciousness of how I feel myself to be. When another person is watching me, she is identifying the state of mind *she* thinks, often quite justifiably, *I* am in. But such identification is all she can do. There is no way in which she can also *have* the subjective state of mind which is private to me, i.e. my inwardness.

Strawson says correctly[77] that self-ascribable and other-ascribable predicates have a single kind of meaning and that the two different ways of ascribing them – one being used when there is self-observation and the other, when the observation is made by another person – are 'perfectly in order'. His discussion, however, is confined to labelled states of mind. His satisfaction with the structure of language would have been diminished, had he considered that the single kind of meaning of the two ascriptions might have been applied to situations which are not as comparable as 'observing myself' and 'observing another person'. To observe oneself is to label and indeed is comparable to any labelling done by another person. Here Strawson's comment is justified. But to *have* a state of mind is a pre-linguistic experience which should not, slapdashly, be assimilated to a predicated and labelled event, no matter how self-ascribed. I prefer Wittgenstein's comment: ' "So you are saying that the word 'pain' really means 'crying'? – On the contrary: the verbal expression of pain replaces crying and does not describe it." '[78] By contrast, Dennett's terminology[79] is not at all helpful. He distinguishes between heterophenomenology (third-person observation) and autophenomenology (first-person observation). This usage creates a misleading impression of symmetry between the two procedures, when in reality the autophenomenological procedure – when it is unlabelled – is self-luminous and therefore *not* a form of observation at all. The rules governing the usage of our language here are glossing over an important difference. And this draws our attention to the inherent inadequacy of labelling by vocabulation: language draws a deceitful veil of gnosis over the subjectivity of qualia by obliging us to assimilate 'I feel' to 'she feels'.

VI

Let me go back to Wittgenstein's aphorism that crying replaces pain, but does not describe it. The operative word is 'replaces'. What is being replaced is not really the ineffable subjective quale in question – as Wittgenstein thought – but the verbal label 'crying' which had been assigned to it for identification. The replacement is important because the label, as we have just seen, is inadequate. The replacement is an act or an image of an act of crying. There is no causal or any other necessary link between the subjective state of mind and 'crying'. The link consists in the fact that we would point to the act or the image of the act or the dream of the act and say:'There! this is how I feel!' For this reason we must say that the act – whatever its precise ontological status as a dream, an image or a real event – is a symbol of the ineffable quale. By replacement, then, we proceed from vocabulation to symbolisation. Events, objects, dreams, images, memory images or images projected into the future as hopes get their symbolic quality by virtue of their relationship to an ineffable occurrence. (We must be cautious: we are used to thinking that a mental event refers to an outside event. But here we get a situation which is the other way round. The subjective, mental event is the unknown. It is the assignation of the outside event which refers to it. The mental event thus becomes the event referred to; and the outside event is the referring event.) Their symbolic character does not consist in a special mode of their representation, occurrence, perception or existence. When the ineffable event is being referred to, the referring event – T. S. Eliot's 'objective correlative' – is called a symbol. The symbol arises when an imaging-consciousness 'gives itself an object'.[80] The *locus classicus* for the definition of 'symbol' is in Mallarmé[81]: 'A symbol comes into being when … step by step, an object is evoked in order to show up the tone of a state of mind (*état d'âme*)'. Whatever their ontological status, they will all do equally well as objective correlatives. The fact that their ontological status is a matter of indifference as far as their symbolic efficacy is concerned has been exploited in France by the so-called *roman nouveau*, in which the narrative proceeds by complete disregard for the ontological status of the events (wishes, perceptions, dreams, memories, phantasies, etc.) strung together. The story nevertheless can be followed and makes sense because whatever the ontological status of the single events, they symbolise coherent sets of subjective feelings and hang together in virtue of their relationships to the states of mind thus symbolised.

Symbols cannot be paraphrased without losing their efficacy as symbols. A paraphrase or any other kind of exegesis would presuppose that we can actually name the symbolised occurrence in some other way, i.e. that it is not ineffable and that the verbal label it carries is an adequate description of it. While mere vocabulation can be extended into symbolisation, symbolisation can never be reduced to vocabulation.

The image, pictorial or descriptive, designates itself as a symbol of the mental event. The notion of designation is important. It indicates that the image does not reflect or mirror or describe the mental event. The mental event is represented symbolically by the image, and the relationship between the inchoate mental event and the image is a symbolic relation. The mental event, we might say, achieves its meaning, its definition, its label through this designation of its symbol. A vague and ill-defined state of sadness or dejection is symbolised, for example, by an image of a weeping willow. The weeping willow becomes phenomenologically its symbol. In poetry and prose literature such symbolisation plays a major role. Since states of mind cannot be described literally, they are referred to symbolically.

'Strether', Henry James tells us in *The Ambassadors*,

> sat there … the confidence that had so gathered for him deepened with the lap of the water, the ripple of the surface, the rustle of the reeds on the opposite bank, the faint diffused coolness and the slight rock of a couple of small boats attached to a rough landing place hard-by.

The confidence, in other words, is a pale and inaccurately literal reference. The real description of Strether's state of mind is symbolised by the words that follow the statement about his confidence.

The designation of a symbol is, first, a hypothesis. Since we have no definition of the mental event by itself, the designation of a symbol can at most be hypothetical. We can never know whether we are hitting, by such a designation, or by the choice of a special symbol, the nail on the head. The head, we recall, is not known; or, rather, it escapes definition. Second, the relationship of meaning which is established between the mental event and its symbol can never be completely fixed. The weeping willow could be used as a symbol. And then again, one could discard it and prefer another symbol. Since the designation of a symbol and the fixing of the specific meaning of the mental event is hypothetical, the fixed meaning could be the 'wrong' or an 'unsuitable' meaning and might have to be exchanged. In making a hypothesis and in designating a symbol, we are finding an objective correlative for the mental event. Only with the help of such an objective correlative does the mental event obtain an effable meaning. It now can be talked about. But what we are thus talking about is not the mental event itself but its objective correlative. Through the objective correlative, the mental event ceases to be wholly subjective and ineffable. It becomes dragged into the public domain – and to that extent it enters the realm of objectivity.

If we now turn to the question of the correctness or suitability or accuracy of these hypothetical designations, we will make a curious discovery. Since the mental event of, say, sadness cannot be labelled other than by the designation of an objective correlative such as, say, the image of a weeping willow, any such hypothetical choice of a symbol must always be either overdetermined or underdetermined. The mental state by itself is ineffable. One cannot look at it and then look at the chosen symbol and then compare the two. The symbol is all we can look at. There is, therefore, no way in which one can suppose that one has made the *right* choice. This does not come from the fact that a certain amount of luck plays an important role. If this were the case, it would be quite clear that, statistically, by the luck of the draw, sometimes one must get it right. The point is that the word 'right' has no meaning here. Since the subjective state is not labelled and cannot be labelled except by the designation of the symbol, there is no meaning in the question whether this symbol is the correct label or not. Therefore, not even by the luck of the draw can one imagine that there are some symbols which are exactly right. The idea of a right symbol in this context is a non-idea.

The objective correlatives that are available as symbols are well-nigh infinite. If one were to arrange them on a scale, one could place the natural images at one end and the non-natural or artificial images at the other and all other images somewhere in between according to the degree of artificiality they exhibit. On the face of it, the distinction between natural and artificial images may appear difficult, even arbitrary. But one can use a perfectly safe criterion for that distinction. We shall call all those images which are compatible and consistent with one another 'natural'; and those images which are not always compatible with one another 'artificial'. In this sense an image of a house and an image of a tree come under the heading 'natural'. The image of the house (house, for short) is an image of something more artificial than the image of a tree (tree, for short). This, at least, is what common usage would suggest. But the house and the tree are perfectly compatible with one another. They both exist in space, or, if one starts with the idea 'image', they both represent something that exists in space and in time, something which obeys the laws of biology and of physics and chemistry. Though one is man-made and the other not, both house and image of house are 'natural'.

At the other end of the scale there are the artificial images which are images of events or things or persons that are not consistent with one another. Among those images there can be a giant, that is, a person who has the size of a tree; or a god, that is, a person who has the strength and power of ten thousand horses, and so forth. The spaces in which these beings exist and operate need not be compatible with one another. One can have an image of oneself in which one is flying with the speed of a space-rocket or an image of one's brother in which he has one ordinary arm and one infinitely large arm or one made up of a thousand swords. With some effort, and theologians and other well-meaning people from all ages have spent a lot of effort in this direction, one can trim these artificial images into a coherent and consistent picture, so that the degree of artificiality which pertains to them is diminished. Thus, for example, the notion of a God who is omnipotent and has created a world in which there is a lot of evil is not consistent with the image of a God who is entirely loving and good. But with some dedication these incompatible images of God can be adjusted and made to appear consistent with each other. This is what theologians are wont to do. The overall set of images, though now internally consistent, would still be incompatible with the set of images which are naturally compatible with one another. Hence the distinction between natural and artificial images remains valid.

When we are choosing natural images as symbols, our choice is limited. It is limited, first of all, by the range of our experience and our knowledge of what is available and second, by the mere fact that the world from which we draw these images is finite. For practical purposes, then, the range of natural images at our disposal is limited. This leads to a curious situation. If we designate a natural image as a symbol of a mental event, we will find that the natural image is not really specific enough to tell us precisely which mental event we are thus intending to symbolise. Take an example. If we choose a weeping willow as an objective correlative, we will find on introspection that that weeping willow symbolically covers quite a range of subjective feelings. If I may play a trick again and pretend that these feelings can be referred to independently, in their own right, as it were, one can say that the weeping willow refers to a feeling of sadness, a feeling of dejection, a feeling of despondency, a feeling of loneliness, a feeling of lonely abandon, a feeling of deprivation ... and probably to a whole lot of other feelings as well. In short, the designation of a natural image is helpful for crawling out of our subjective isolation; but it does not do the job with any great precision. Clearly here natural symbols underdetermine the states of mind they symbolise.

One can stand this imprecise relation on its head by making the symbol one chooses less natural and more artificial. One can take the weeping willow tree and make it larger than it is or can conceivably be. One can place it in a lunar landscape and have it inhabited by fish which have human voices. The symbol is thus made more specific than any willow tree we might have encountered by the bank of a river. In being more specific, it has an increased ability to symbolise, and, if genuinely specific, it will now symbolise or label not a whole range of feelings in the area of sadness and depression, but one very specific state of feeling. We now have a very high level of specific meaning of one single feeling-state. However, as soon as this is achieved, we make another curious discovery. We find that that same feeling-state – a very special kind of depressiveness clearly symbolised by an over-life-size weeping willow tree in a lunar landscape populated by fish with human voices (there is no other way of referring to that feeling-state) – can be symbolised with equal precision by a large number of other symbols. It is by no means wedded to the weeping willow in a lunar landscape. On the contrary, any large tree in any non-terrestrial landscape possibly populated by other strange creatures will do equally well and will be equally precise. Thus, for example, a state of numinous feeling can be labelled precisely by the image of Jehovah, by the image of finding oneself in the depth of a dark cave or by finding oneself entering Chartres Cathedral on a winter's afternoon when the sun is setting behind the stained glass windows at the west end. There is no lack of precision now. But there is a variety of choices open to us, each one with the same high level of precision as far as the feeling-state is concerned. Here, then, we have a case in which the mental event is overdetermined by a whole range of symbols.

One and the same objective correlative can symbolise a whole range of mental events which remain, therefore, underdetermined; and one mental event can be symbolised by a whole range of objective correlatives so that it is overdetermined. In stressing the plasticity of the relationship between the symbols or objective correlatives which bestow meaning and definition on mental events and the mental events themselves, I am not saying anything very new. I am merely extending to normal life something with which psychiatry has been familiar for a long time.

I believe that Freud was the first to draw explicit attention to this phenomenon of plasticity, though we need not necessarily agree with his interpretation of it. In his essay 'A Child is Being Beaten',[82] he wrote of the phantasy (= image or, better, persistent image or perception) that a child is being whipped on its bottom. Even those of his patients, he wrote, who freely admitted that they had this phantasy had great difficulty in telling exactly what the specific meanings of the phantasy were. That is to say, they had in their mind's eye the objective correlative image of a child being beaten. But they were unable to say with any degree of certainty whether this symbol meant a feeling of indignation that the child was being beaten; or a feeling of sadistic pleasure in seeing a child being beaten; or a feeling of excited anticipation at the prospect that a child would be beaten; or a feeling of power in thinking that a child was about to be beaten; or a feeling of hope that the child would be beaten; and so forth. They were able to suppose that the symbol referred to a large number of qualia and were unable to decide with any precision which of the qualia in that wide spectrum was the one intended by the objective correlative.

The image itself, in other words, was lacking in definition. All the same, though lacking in definition, the phantasy kept recurring, and we are therefore obliged to believe that it was used, in spite of the absence of decisive delineation, to bestow meaning and definition on a mental event. But the mental event thus symbolised or meant by the symbol was underdetermined. If one pushes the analysis of the image further, one will see that its very lack of specific lineaments shows that it is very much what we have called a natural image; and it is because of this simple naturalness that it covers such a wide spectrum of mental events. It bestows meaning equally on those mental events which, if this were possible, one would describe literally as 'pleasure', 'fear', 'hope', 'apprehension', 'wish', etc. All of these states of mind are very different, and one would have to look towards a more specific objective correlative in order to be able to say with any degree of certainty which quality in the wide spectrum is meant. Clearly, the image would become more specific in its definition of a mental event if it were made less natural: that is, for example, if the beating were being done by a giant, a demon or a god. One would then be able to make out better whether the beating was pleasurable or painful, feared or desired. As long as the beating is being done by, say, the father, the ambiguity remains: it could symbolise pleasure in being beaten as soon as pain or apprehension. But when the person who is doing the beating becomes more strongly defined, the range of mental events covered by the objective correlative becomes narrower.

If here we had an example of how a wide range of mental events can be covered by one image, the condition known in psychiatry as *algolagnia* provides an example of how one mental event can be symbolised by a wide variety of quite different, though related, objective correlatives. The mental event known as *algolagnia,* if it could be described literally, would be defined as a mental state in which pleasurable sexual excitations are mixed with the occurrence of terrifying events. In the possible and conceivable objective correlative it makes no difference whether the mixing of pleasure and terror is accidental or contrived; and the terrifying events may be of various kinds. They may involve the deaths of persons or of animals or merely the threat of death. They can involve the destruction of health or property or the total annihilation of material goods. The destructive acts may be performed, in the image, by the person himself or herself, or the person may induce somebody else to perform them, or he or she may remain a mere spectator. They may be willed by God or by a devil. No matter how precisely the correlative image is formed by a departure from a mere natural event, all of these quite different images will converge upon one and the same state of feeling.

Thus we find a situation which can be illustrated by the following diagram:

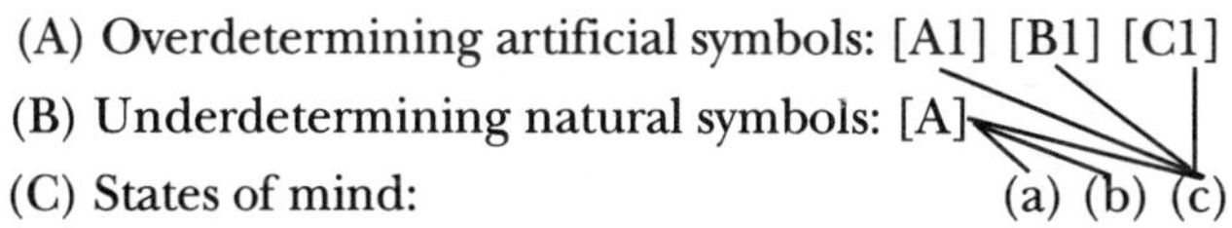

Figure 1

The states of mind (C) are either symbolically underdetermined or symbolically overdetermined. Either one gets too many artificial symbols (A) referring to one state of mind; or one gets too many states of mind being referred to by one single natural symbol (B). A natural symbol leaves the state of mind underdetermined because it does not make it unequivocally clear which specific state of mind it refers to. An artificial symbol leaves the state of mind overdetermined, because the same state of mind can be meant by a large variety of different symbols. This is exactly what one would expect. Since a state of mind cannot be identified as it occurs, it would be not only surprising but also inconceivable if any one chosen symbol should hit the nail on the head. In fact, it is impossible to say what 'hitting the nail on the head' might mean, because the state of mind cannot be identified with precision independently of its symbol. Even if one goes by the verbal label assigned to it, precision is absent; for a label is, as we have seen, a mere hypothesis. There is either a lack of precision in the *relatum*, as when a weeping willow tree symbolises sadness, despondency, loss, nostalgia, etc.; or uncertainty in the choice of referent, as when Jehovah, God the Father, the Buddha, Jesus, a demon, a giant, etc. can be used to symbolise a state of numinosity.

VII

Looking back upon both vocabulation and symbolisation, we conclude that language always plays crucial parts. But it will not help to think that it is always the same part and that that part is an 'interface between knower and known'.[83] On the contrary, none of the parts it plays can be considered such an interface. When we are trying to introspect, language lets us down because it forces us to report our findings in a way which does not do justice to self-luminosity, for it forces us to say 'I am conscious *of* something', which is an expression which does not cover the case. When we try to provide a label for a state of mind it fails again, for a different reason, because it does not allow us to distinguish between me saying 'I am sad' and another person saying, about me, 'he is sad'. When we turn to symbolisation, the employment of language becomes more complicated, though it may seem a little more satisfactory. A natural symbol can be described in so far as anything can be described. There is no real problem here with language. The problem of precise reference lies in the fact that we are now symbolising the state of mind rather than providing clear reference to it. When we come to artificial symbols, they too can be described, though for a different reason. Every artificial symbol is an assembly of elements wrested from their original position in something that could be used as a natural symbol. What makes the artificial symbol artificial is the mode of the assembly of these disparate elements, not the character of the separate elements themselves. This means that the artificial symbol, though we have put it together and though there is no known place in the world where it might be found, can, nevertheless, be described verbally. It is put together by loosening distinct elements of the natural world from the position they naturally occupy and then reshuffling them to present a totally new image, one which cannot be found in the natural world in that precise form. The highest flights of imaginative creativity never can reach beyond such loosening and such reshuffling.

People who are concerned with religion and the supernatural have always benefited from the fact that there is no image of a Godhead, no miraculous event they can think of, which cannot be depicted verbally; or, at least, which cannot be depicted as verbally as any other event. Opponents of religion and the supernatural have always barked up the wrong tree because they have used this fact to demonstrate that all presentations of the supernatural are anthropomorphic and have concluded from this demonstration that claims that supernatural events can take place must be false. But there is nothing surprising in such anthropomorphic representations. The only surprising thing would be if one could come up with a non-anthropomorphic representation! Language does play a crucial role, because we are obliged to use it and we are using it as best we can. But we would be committing grievous errors if we allowed ourselves to be guided either lexically or normatively by its rules and its limitations. The Wittgensteinian prescription that we should use it either 'correctly' or not at all is not helpful.

Looking beyond language and its constraints, we are now more aware than we were at the beginning of this chapter that consciousness cannot be one of the terms in the relationship which we know knowledge to be. Wherever we have looked into consciousness, we have been able to explain how one can break out of silent inwardness. But we have also seen that every such attempt to break out leads to situations in which there is no one-to-one correspondence between the media we use to break out and the inwardness we are breaking out from. When we reach inward, we remain silent. When we reach backward, the articulated labels we assign to states of mind are hypothetical. When we reach forward and designate symbols of states of mind, we first meet natural symbols which are imprecise because any one bestows meaning on a variety of states of mind; and when we proceed to artificial symbols, they remain uncertain because any one state of mind can be symbolised by a variety of related symbols. Whether our inner state of how we feel ourselves to be is Kierkegaardian or Nagelian, a one-to-one correspondence between how we feel ourselves to be and how we express it and drag it into the articulated domain, public or not so public, is unobtainable.

The relationship between silent consciousness, vocabulised or symbolised, as the case may be, and neuronal circuitry amounts to some kind of dualism of mind and body, even though the silent part of consciousness must be seen to stand in a causal relation to the physical activities of the body. The elements of mind that are made physically visible by the labels or the symbols that get hypothetically or uncertainly attached to them stand in a tenuous relation to the states of consciousness. Since they do not exhibit any kind of one-to-one correspondence to the mental events they label or symbolise, one must think of dualism. For though labels and symbols are visible and can be referred to events which take place in space and time, their relations to mental events is always overdetermined or underdetermined and in any event always hypothetical, so that one cannot envisage a definable relationship between the mental events that are being dragged into the public domain as labels or symbols and the mental events themselves.

There are indeed two different detection chains which never intersect. On one side there are physical events. One can track them along the causal chains they form, into the body, inside the body and then outside the body again. Certain olfactory inputs generate nerve reflexes which cause adrenalin to flow, and that adrenalin can be detected in the blood, put into test-tubes and related to the behaviour of a measuring instrument. Many of the causal chains are very well known. But at no point on such a chain does a mental event ever put in an appearance. The word 'fear' used to label the inchoate and ineffable state of consciousness generated by this chain of physical events is nothing more than a hypothetical label, and it is only custom and conditioning and lack of attention which attaches it so intimately to the state of mind that text-books are in the slapdash habit of saying that the sight of a wild lion causes fear, that fear generates adrenalin and that adrenalin can be measured. In reality, the sight of a lion causes, at one end of the chain, an inchoate stirring of an aura of consciousness. It does not 'cause' the identified state of mind we know as 'fear'.

The other chain of detection, which starts with introspection of a subjective state of mind, always leads to other states of mind, regardless of whether they are more or less labelled or symbolised. The spiritual or pneumatic appearance of these states is, as was noted above, really noticeable. By looking into one's subjective feeling oneself to be something – which is a state of self-luminosity – one can very well notice how one feeling oneself to be changes either suddenly or gradually into another feeling oneself to be. But one will never encounter, if one is following this road, a physical condition or event. Introspection can never lead to anything sufficiently specific that could be called a physical event. By the time the result of introspection is labelled or symbolised, the state of mind then encountered and potentially visible or physical stands only in a hypothetical relation to the initial mental event and so cannot be taken to stand in a causal relation to it. That is to say, one's consciousness travels along paths of qualia, and on this road one will never find anything but qualia. Thus there is a lot to be said for the image of the ghost in the machine – except that on this view, the ghost is helplessly imprisoned in the machine and cannot direct it and the machine, though holding the ghost captive, cannot inflict anything more tangible on the ghost than broadly classifiable qualia.

With the possible exceptions of Jesus Christ and Richard Wagner, there is no subject in the world which has been written about so much as the relationship of mind to body. Philosophy is replete with theories of materialism, spiritualism, idealism, eliminative materialism, stable-state materialism, interactionist dualism, predetermined harmony parallelism and identity theories, to mention the most widely debated ones. This is not the place to discuss and weigh these theories, but rather a place to draw attention to the fact that the inability of the two detection chains to intersect provides an opportunity – paradoxical though this may sound – to explain why and how interaction takes place.

There are countless observations which tell of interaction or, better, which call for an explanation in terms of interaction. The formation of an intention to perform an action is a mental event. The action which follows is a physical event. Attempts have been made to interpret this interaction as a result of an intersection of the two detection chains. Thus, for example, Popper and Eccles hold that the formation of an intention *leads* to a 'wide-ranging negative potential over the top of the brain' which builds up for almost one second.[84] Pribram accounts for conscious voluntary control in terms of 'a change from feedback to feedforward organisation of the nervous system as a whole' and sees voluntary effort as a 'necessary concomitant of openloop processes'.[85] For Eccles and Popper and for Pribram the operative concepts are 'leads to' and 'necessary concomitant'. They are designed to gloss over the real problem of interaction or intersection by creating the appearance of a gliding causality at work. R. W. Sperry is even more categorical, though much less specific. He writes that 'as high-level dynamic entities, mental processes control their component biophysical, molecular, atomic and other subelements'.[86] Sperry wants the reader to see his emphasis on the word 'component', because thus one gets the impression that there is one single process involving mental and physical events and that that process simply has several different 'components'. But the real catchword is 'control', not 'component'. And the use of that word is designed to perform the same gliding function as is performed by Eccles' 'leads' and Pribram's 'concomitant'.

It is widely believed that a breakthrough to understanding the intersection of the two detection chains was made when the neuronal concomitants of mental events were discovered and were used to replace the external objective correlative and other labels we have used so far. Thus it was held that fear is correlated with 'seeing a lion'. Now we think we can be more precise because we have found the neuronal accompaniment of fear, and thus it is thought that a science of subjective feelings will be possible because we can see exactly how mind and body interact when adrenalin is produced. But this project is a delusion. There is no gain in moving from an objective correlative or a mere label to an internal, neural concomitant. A silent neural event, no matter how much adrenalin is contained in it or produced by it, is not the same as 'feeling fear'. The lack of one-to-one correspondence remains and the two detection chains do not intersect. In a way this conclusion is inevitable because of the two modes of detection. In one chain there lie all those events we are aware of; and in the other chain, there lie all those events we diagnose and measure in terms of the results they have on other events.

These gliding and glossing strategies are supposed to help us to hold fast to the uniform action of causality which prevails in the dualism between mental and physical events without turning the one into the other as so many other theories are wont to do. But the gliding from mind to body cannot conceal the fact that there is a glossing over of difficulties. I would like to suggest that the difficulty can be dealt with more satisfactorily by abandoning the belief that there are uniform causal connections by which mental events interact with bodily events. If, on the contrary, one supposes that there is no intersection and no one-to-one correspondence, one can use this absence of clear causal relationships and the presence of hypothetical labels and hypothetical symbols to show how mental and bodily events may interact or influence one another. I would claim that one can explain the fact that the mental event 'I want to raise my right arm' makes my right arm rise not only without supposing that there is a causal relation between mind and body, but precisely because there is *no* such relationship and because the relationship is hypothetical rather than causal.

I suggest that the process from a (defined) mental event such as 'I want to raise my arm' to a neuronal-physical motor event that my arm is rising goes *via* inchoate consciousness. This suggestion is not a version of causal interaction, either in the original Cartesian or the modern Ecclesian sense. On the contrary, it amounts to the suggestion that intersection takes place *because* there cannot be causal interaction.

The inchoate state is something like a shunting yard. It is generated neuronally and, since it is inchoate, can be variously interpreted. Any hypothetical label that can be assigned to it and the image or event which corresponds to that label can be understood as the 'effect' of the physical processes which generated the inchoate state. The reverse process can also take place. If I will something, such willing can be a label which I stamp on an ill-defined inchoate state. The stamping gives it a more specific, silent shape, and *that* shape initiates a chain of neuronal events. It is well known that many different mental events can cause identical physical reactions. One can raise one's arm for many different mental reasons. A specific mental event can act as a 'cause' of a physical event because it is at first scrambled into something less specific (i.e. it is being stamped upon inchoate consciousness) and more crude than an articulate expression. And from that condition it activates specific neuronal circuits. In seeing inchoate consciousness as the interface, we can also understand why the link between mental events and physical events is usually much looser than simple causality would lead one to expect.

In this way one can also explain the comparative successes of manipulative psychology and possible therapies. Paradoxically, these successes do not come from the fact that we understand the causal connection between behaviour and mental states; but from the fact that there are no such causal connections. Given the fact that any identification of a mental event is purely hypothetical, it is possible to shift the initial hypothesis to a hypothesis which is socially or emotionally more desirable or more useful. And given the fact that the mental event which emerges from a neural condition is unlabelled and thus fairly open-ended, with its edges frayed, it becomes tolerant of a large variety of hypothetical specifications. In this way one can see how mental events are flexible and, because of this flexibility, manipulable. What is being manipulated here is the frayed nature of the inchoate stirring – the qualia. If the qualia were fully definable and less frayed at the edges, such manipulation could not take place. Once manipulated, these mental states will and can have a causal influence on other physical events inside our body. For the link between the physical events inside our body and the inchoate stirrings which we cannot pin-point or identify is, though literally indescribable, causal. Paradoxically, it is the very tenuousness of the causal link between identified mental events and bodily events which makes it possible to explain and account for the fact that there is a link between the undefined mental events and the body.

VIII

This view of inchoate mental events and their, of necessity, loose connections with any labels and symbols which define and specify them has very damaging implications for three widely patronised projects in the philosophy of mind – for AI, psychoanalysis and the functionalism of cognitive science.

Artificial Intelligence

For decades, a great many people have been intoxicated by the notion that it will soon be possible to program computers to perform as well as human brains/minds, and it is widely believed that the only question which remains is whether such computers will actually be able to be conscious. Some arguments have been advanced against such hopes; but for the most part, it has been overlooked that human brains/minds do not perform all that well and that if there are to be brain/mind simulators, one would, first of all, have to expect simulations of the incessant mistakes, misidentifications, hypothetical guesses and symbolic ascriptions human minds are 'capable' of. In a way, we are talking here about the deficiencies of the human brain/mind, not of its powers. However this may be, it cannot go without notice that recently the voices of such great experts in computer facility as Gerald Edelman[87] and Roger Penrose[88] have been raised almost simultaneously against these hopes. For good or ill, the human mind can do things a computer cannot be programmed to do because, Edelman says, the nervous system is not a fixed set of connections. Its repertoire is 'degenerate', so that some non-isomorphic groups must be isofunctional: nothing is ever precisely repeated or reproduced. Penrose, looking from a different perspective, suggests that the brain/mind cannot function like a computer because the new ideas it gets – true ones as well as false ones – come suddenly as wholes; not as results of logical steps.

There is one feature which AI, psychoanalysis and functionalism have in common. In all three systems it is essential that any given mental event be capable of an ultimate and ultimately unequivocal definition. In the theory of consciousness presented here, there can be no such unequivocal definition of mental events: the relationship between inchoate mental events and their labels or symbols can never be a one-to-one correspondence. This is clear enough in the case of symbols, regardless of whether they are natural or artificial ones. Labels can be mistaken more readily for descriptions than symbols. But labels too, in the absence of an ability to refer to mental states independently and other than by their verbal labels, are shown not to be descriptions; and are shown to be almost as much like objective correlatives as symbols themselves.

All projects of AI are bedevilled by the fact that we have no idea what really would count as 'intelligent' behaviour, and it is therefore impossible to tell whether a machine is simulating intelligent behaviour or not. Whatever machines can do, we cannot know whether they behave 'intelligently' or not, because we do not know when the term 'intelligent' is applicable. But let us, for the sake of argument, leave this fundamental problem aside. There remains the fundamental objection to AI, brought up by H. L. Dreyfus in anticipation of the arguments of Edelman and Penrose, as it were,[89] that, among other things, human intelligence operates on the fringes of consciousness and involves a very high, perhaps an almost complete, tolerance of ambiguity. Dreyfus therefore concludes that no machine can be programmed to behave 'intelligently'. This argument is entirely correct but, in view of the present theory of essential ambiguity and even 'poliguity', does not go to the heart of the matter. For in so far as we know which ambiguities our minds can tolerate, we can certainly program a computer to tolerate those same ambiguities. The real difficulty is not that we can tolerate ambiguities but that, given the inchoateness of mental events and the absence of one-to-one correspondence between label or symbol and mental event, we can never gauge the *extent* of that ambiguity. In order to program a computer to be as tolerant of ambiguity as we are, we must know not only that we are tolerant but also the exact extent, i.e. the limits, of our tolerance. This is something we cannot possibly know, for if we did, there would not be all that much ambiguity. Since there can be no such limits, it follows that no machine can be programmed to imitate a human mind. This is not due to the fact that the human mind is too complex. Machines can most probably be made equally complex. It is due to the fact that the human neuronal machine's complexity produces an unlimited high level of ambiguity. It is that ambiguity and not the sheer complexity of the neuronal machine which lies at the heart of our problem-solving ability.

AI enthusiasts have always made it too easy for themselves.[90] They believe, to a man and woman, that problem solving comes from the removal of ambiguities. But it has turned out that the human neuronal machine's ability to solve problems has a lot to do with its infinite capacity to generate ambiguity. This means that the more intelligent computers become, the less will they be like the human mind. A *truly* intelligent computer would be a computer which randomly malfunctioned all the time. Only such random malfunctioning would be comparable – though by no means identical – with inchoate and fuzzy states of mind. The project of AI is misconceived on the assumption that what we are aiming at is greater and greater precision and more and more removal of ambiguity. In reality, because of the plasticity of the relationship between inchoate mental events and their objective correlatives, an artificially intelligent machine would have to produce a comparable level of non-specific and undefined awareness. The project of AI is based on a mistaken view of mental events. This is no criticism of programmers and cannot hope to be remedied once we *really* know human minds. Indeed this conclusion does not depend on our, perhaps only temporary, ignorance of human minds. On the contrary, it follows directly from our *knowledge* of human minds as inchoate and as standing in nothing but hypothetical relationships to labels as well as to possible symbols. In short: any computer program must reflect the programmer's labelled state of mind and can be put in place only after a lot of initial, hypothetical groundwork has been done. A computer cannot be programmed to carry out the inchoate states of mind and their interconnections – for the simple reason that a programmer cannot know what they are. Hence, the machine cannot imitate, cannot even simulate, the programmer. It can only be made to simulate the hypothesis the programmer has made about his or her states of mind.

Psychoanalysis

Initially, the Freudian project of psychoanalysis looked more promising. For the project was based on the clear understanding that the identification or labelling of a mental event in our conscious mind is questionable and, at best, provisional. This was a great step in the right direction. Behind such provisional designations of labels, Freud argued, there lies the unconscious. Things are not what they seem. Freud nevertheless held fast to the idea that each mental event is, in the last analysis (= *psycho*analysis), identifiable as one rather than another 'given'. Indeed, he surmised that, since he had found that in our conscious mind, our motivations are not what they seem, we must be motivated by our unconscious mind rather than by our conscious mind. Rather than move in the direction of uncertainty and believe that the conscious mind is forever plastic, Freud decided to hold fast to the notions of causal one-to-one correspondence and to determinism, i.e. to an ultimate certainty, and assumed, therefore, that our motivations must come from the *un*conscious. Though Freud never put it in this way, the strongest reason for postulating the unconscious is that it allows us to think in terms of sets of causally determined relations between experiences and responses which are so lamentably absent from the conscious mind. The contents of the unconscious mind, Freud believed, are hard and fast, and in order to get away from the loose and hypothetical connections that are made in our conscious minds, we must dive into the unconscious mind. The unconscious had to be postulated in order to save those appearances of determinism to which the conscious mind does not appear to be subject.

Even if we leave aside the whole question whether there is an unconscious mind and whether its contents are contents that have been repressed, Freud's argument runs up against the obstacle of the absence of one-to-one correspondence. If there is no hard and fast relationship between mental events and labels or other objective correlatives, one cannot define the meaning of a mental event in any absolute sense. This goes for the unconscious as much as for the conscious. Take an example. Suppose somebody is conscious of loving his country. One has no difficulty in probing and in showing that that 'love of country' is a symbol which stands for a mental event which could also be symbolised in a different way. Our theory would lead to the view that there is a whole range of optional symbols and that they are all, more or less, equivalent in their power to label a given mental event. Not so Freud. Though our doubt as to the finality of 'love of country' is shared by him, he would suggest instead that we probe and find out what the *real* label for that mental event is. That *real* label, Freud would say, lies in our unconscious and has to be unveiled. In our unconscious mind we would then find that the correct label for the mental event in question is 'love of the mother'. Freud would thus concede, and this was indeed one of his greatest discoveries, that the consciously proffered label is very dubious and certainly not final. But he would not concede that the label hidden away in our unconscious mind is yet another hypothesis and that the two hypotheses, the conscious one and the unconscious one, are interchangeable and both *equally* non-final. The most he would concede is that, if even an unconscious label turned out not to be ultimately final and turned out to be unable to tell us the really literal meaning of the mental event, this was because the psychoanalysis had not been carried sufficiently far. At the end of the tunnel, Freud fervently believed, there always had to be light. This belief is rejected by the present theory of the relationship between mental events and their labels.

Functionalism

Last we come to the current theory of functionalism. This theory is a very welcome addition to the arsenal of theories about the relationship between mind and body. It is specially relevant here because, like the present discussion, it takes mental events very seriously. Functionalism is, unlike many other theories on the subject, not concerned with reducing either mind to body or body to mind. It is also a refinement of ordinary dualism because it can envisage a way in which mental events and bodily events are laced together by being parts of a functional system. But this is, as we shall see, where the rub lies. Functionalism 'construes the concept of causal role in such a way that a mental state can be defined by its causal relations to other mental states'.[91] Functionalism's great attraction is that it is neutral about whether mental states are *really* mental or physical. All it is concerned with is the individuation of states by reference to their causal role.

Functionalism was put forward because it was suspected that there are empirical generalisations about mental states which cannot be formulated in the vocabulary of neurological or physical theories. The view was therefore advanced that behaviour of man or machine is not explained by the physics and the chemistry of its computing mechanism, but by the mechanism's program.[92] In this view, a mental state does not have to be defined or described in its own right, in isolation, as it were. It is simply functionally part of an entire system, which contains, also as functioning elements, non-mental events. Given such functioning systems, it is then held that mental events, which means the qualities of mental imagery, such as colours, are not to be taken as properties of private, subjectively internal states (qualia), but as functional properties of objects which, when observed and used as colour words, will prompt the subject to say that the object had the colour in question. In this way, and without further ado, functional properties can be simply threaded into the chain of functions.

The viability, not to say plausibility, of these arguments depends on the assumption that in all cases mental events can be identified and labelled. Once they are named, they can be threaded into the functioning system, and can even be seen to be so threaded, at the appropriate point. The trouble is, as we have seen, that no mental event carries its own label or symbol and that the procedure that is required for tagging it always leads to a situation in which any tag is tentative and capable of making way for an alternative tag. Functionalists always deceive themselves by mistaking the symbol or label for the mental event. It is no surprise that a large proportion of philosophers guilty of the 'slapdash' assumption discussed and criticised above call themselves 'functionalists'.

If one perceives, for example, an enduring sound, one can easily delude oneself that the enduring sound *is* the mental event roughly referred to as 'perception of enduring sound'. If one allows that delusion to persist, the mental event is, falsely, taken to have come with a final label and can even be described as a 'functional theoretical event'.[93] In that case and as such, there is no reason why it should not be threaded into a functional system as it stands. The point, however, is that it does not come with such a final label. Or consider the famous Chinese Room argument put forward by John Searle. The man in the room, Searle insists, may well function as if he actually *understood* Chinese, but he most certainly does not *really* understand Chinese. I consider Searle's argument to be logically impeccable. The reflection is based on the difference between the subjective understanding of Chinese and the simulation of a behaviour pattern. It is parallel to the problem involved in colour or sound perception. One can most certainly learn how to use the expression 'red' and thus learn to communicate with other people who have also learnt to use it. None of this proves, however, that any of the people involved in this speech exercise have the same sort of subjective awareness when they employ the word 'red'. Functionalism thus appears viable because one can easily have the delusion that, since one knows how to use a word or label a mental event, one 'knows' what it feels like to experience the mental event or what it feels like for the mental event to occur. Nevertheless, the easy equation of 'knowing how to use' and of 'feeling what it is like to have' is a delusion.

Or consider functionalism as it appears in Fodor's *Language of Thought.*[94] To have a belief (= propositional attitude) is to stand in a computational relation to an inner representation. An agent, it is held, has something in his or her mind such as 'Jim thought he saw a bird', and it is that mental content, not the fact that there was a bird on the bough, which functionally explains Jim's behaviour of firing in a certain direction.[95] But it is a mistake to attribute a semantic property to the internal states of a computer, and it does not matter whether that computer is an Apple Macintosh or a human organism. Fodor believes that computing devices are capable of such semantic appropriations, and these appropriations are called Mentalese – an inner, mental language. But in all computing devices, artificial or human, the semantic properties are not inherent but are inserted by a programmer. It is that programmer, not the Mentalese by itself, which establishes the link between the machine and the world. Fodor, of course, has suggested a theory as to how the Mentalese establishes itself inside the device without being put there on purpose by a programmer. In his *The Modularity of Mind,*[96] he says that perception is the computation of a series of increasingly abstract mental representations of the environment which proceeds from peripheral stimulations to full-blooded conscious percepts. But this is not an acceptable description of what takes place. In reality there is discontinuity and a sharp break when the ineffable inchoateness, produced as the last link in a neuronal chain, stops and verbal labelling (attribution of semantic property) begins. The articulated Mentalese, which is what Fodor's internal representations amount to, is not the product of neuronal circuits. At most, the neuronal circuits provide the raw material of inchoate mentality which is eventually sculpted into articulated image or sentence or sentence about an image. The mental event by itself, which stands at the end of a neuronal circuit, has no such effable properties. For this reason, no functionalism, even the functionalism of the Fodor variety, can be viable. More's the pity, because a theory which accounts for mental events without reducing them to non-mental events would be precisely what is required.

Daniel Dennett has proposed a somewhat different view. He suggests that we start with mental representations that understand themselves and hold fast to them. This self-understanding, he suggests, should not be taken to be due to a homunculus inside our body inspecting the representations. If it were, then we ought to postulate that the first homunculus is being inspected by a second, and the second, by a third, and so forth. Dennett suggests that we rid ourselves of such infinite regress by thinking instead of the first homunculus as being replaced first by a committee of very simple-minded homunculi and then of that committee being replaced by yet more stupid homunculi, and so on until we get the operation performed by a whole army of homunculi who are so stupid that they cannot even talk. So it is homunculi all the way down until we strike whole armies of idiots who perform without murmur. This image is as charming and as nonsensical as the explanation that the universe rests on a turtle and that that turtle rests on another and that it is 'turtles all the way down'. It is not easy to know where it gets its charm from; but the nonsense derives from the fact that though one cannot start with representations that understand themselves (= labelled states of mind), there is nothing in the labels to direct us to the right homuncular idiots at the bottom who will perform precisely the task of generating the specified representation. The reason is obvious: the representation which understands itself has two different sources. One part, the label, comes from the language we speak, and it is this language which gives the representation a semantic wrapping. The other part, the inchoate state of mind, does come from the work of those homuncular idiots, but that work remains as inchoate as the poor mental calibre of those idiots would lead one to expect. In any case, the semantic wrapping is only a hypothetical attribution and cannot indicate where to look for the silent idiot army which will perform the task and produce the content capable of being wrapped in a wrapper in virtue of which we could say that the representation understands itself.

On the view of mental event argued above, the mental representations which understand themselves are not just self-luminous qualia, but fully fledged verbally labelled or symbolised states of mind. From such labels or symbols there can be no one-to-one correspondence back to committees of homunculi, because these idiot workers are not gifted with semantic intentions but are labouring in deadly silence, and no amount of their silent sweat can produce anything more than obscure intimations of vaguely identifiable mental qualia; and certainly no representations that understand themselves. They are, after all, idiots, not mental poltergeister[97] miraculously pounding on synapses to transform their operations into verbally defined mental states.

IX

Dennett's homuncular reduction suffers from the flaw from which all AI attempts and all psychoanalysis are suffering – that is, from the mistaken belief that states of mind are unequivocally identifiable. They are not; and this is the chief reason why consciousness cannot mirror the world and why it cannot be equated with the 'knower' in the cognitive relationship of knower to known. It is rather the other way round. The natural and artificial images the mind designates and the images indicated in the labels given to states of mind mirror the mind. This is something like an ultimate philosophical inversion: some aspects of the world can be used to mirror the mind. The mind cannot mirror nature; and, what is more, cannot be *made* to do so. But through the labels we assign to mental states and especially through the symbols we designate to refer to mental states, nature can be made to mirror mind. Consciousness or mind is not a facility for receiving or storing information about the world – even though it is, in part, generated by information received by the organism. But such generation always results in nothing more than silent stirrings. The interpretation or fixation of the meanings of these stirrings is hypothetical and, when symbols are used, either overdetermined or underdetermined. One can never get it right, for the very simple reason that the relation has to be *symbolic* because the state of mind in question cannot be referred to literally with any adequacy. And such literal labels as we can attach to them may well be wrong because, again, we cannot focus on the state of mind by itself and are incapable of testing whether the label is correct or not. Label or symbol – this is all we have: *stat rosa pristina nomine, et nomina nuda tenemus.* The nature of mental events is such that there are three possibilities:

1. They remain encased in the body as pure subjectivity of inchoate, ineffable, self-luminous feeling-states.
2. They are subjected hypothetically to a conventional vocabulary.
3. They are expanded into symbols.

If they are subjected hypothetically to a conventional vocabulary, there is a temptation to forget about the hypotheticalness and to commit a naturalistic fallacy by thinking that they *are* as the labels allege them to be and that in so being, they are causally related to the rest of the body as well as to the world. If they are expanded into symbols, there is a temptation to commit either a similar naturalistic fallacy in forgetting that natural symbols underdetermine them; or a supernaturalistic fallacy, in forgetting that denaturalised symbols overdetermine them. Needless to say, all of these temptations ought to be resisted.

2

THE DUBIOUS CREDENTIALS OF POSITIVISM

We have established that knowledge is not a relation between mind and matter, that is, between consciousness, evolution's most astonishing production, and the rest of the world. When we started with introspection, we got consciousness, and with consciousness, we got fashioned images – natural ones or artificial ones. But however we turned and no matter how we manipulated those images, we never got beyond consciousness. Any relationship was a relation between consciousness and its images. If there was cognition, it was cognition of images; not of the world. Alternatively, we found that consciousness can be labelled by words or sentences. But such labelling is hypothetical and the determination of the labels comes from the language one is speaking – not from the constitution of the world or its properties. Labels do not lead to knowledge, and the relationship they establish between knower and known is a social, not a cognitive relationship.

Let us therefore explore what happens if we start at the other end – not with the knower, but with the known; or, more correctly, with what is supposed to become known – the world. If we take the world for granted, we can assume that the cognitive relationship results from the fact that the world stamps its image upon the knower or instructs him or her to know. This will mean that the knower is a passive recipient, and the more passively he or she receives instructions, the more correct the knowledge that is left with him or her. Now look at the other side of the relation of knowledge. Knowledge, we said, is a cognitive relationship between an inner and an outer world, between the inside of a living cell and the rest of the world. The inner and the outer are separated by a skin or a membrane, and the cognitive feature of the relationship consists in the fact that the membrane knows what to let through. It is a perfectly reasonable supposition that the membrane's knowledge results from the fact that the outside world has instructed the cell and its membrane and that the cell, to receive the instructions correctly, has listened as passively and attentively as it can.

When Wittgenstein remarked that there is nothing specially 'mystical' in the way *how* the world is, he could not have sounded so persuasive had he not stood at the end of a very long tradition which had maintained, in one way or another, that we find out how the world is by letting the world instruct us. Provided we place ourselves in an untrammelled causal relationship to the world and allow it to pound upon us without hindrance, it will stamp itself, so to speak, upon us and, thus, we will have knowledge. Obviously, Wittgenstein was able to convince himself as well as his readers that there is nothing mysterious or mystical in the results. The world, so to speak, is a battering ram and we are its willing victims. Provided we do not put up defences, it will batter us into a state of knowledge.

In this view any anthropocentric predicament is automatically brushed aside. For in this view 'having knowledge' is taken to be like 'being a victim'. There is not even an ontological circularity which arises when one has to think up a method by which one can know the world. One can only think up a correct method if one knows what the world is like. But since one does not know what it is like, one cannot think up a correct method of knowing it. If, on the other hand, the knower is taken to be a passive victim who receives instructions, the knower does not have to involve himself or herself in such a circular predicament. The totally defenceless victim who neither puts up a hindrance nor takes evasive action is a knower who has delivered himself or herself of all idols and therefore has lost, so to speak, all personal attributes and all specific identity. The knowledge he or she is left with is therefore the knowledge of a battered victim. The more defenceless the victim, the more objective his or her appreciation of the nature of the suffering inflicted on him or her. The knowledge thus enjoyed by the battered victim is, in its total passivity, almost identical to the knowledge an observer who is standing on an alien peak might have. The total victim and the transcendental spectator have one thing in common. They may both suppose to see the world as it is in itself.

The broad line of this tradition stretches, conventionally, from Bacon to Locke; and from Locke, via Comte's positivism to J. S. Mill and to the Vienna Circle and, finally, to the so-called Received View. For brevity's sake, I will refer to the whole tradition as 'positivism', even though this is strictly speaking not correct. Positivism as such was an invention of Auguste Comte, who had adopted the term to describe a number of qualities: 'positive' knowledge had to be real, useful, certain and precise. Knowledge, he said, had evolved from a theological stage to a metaphysical stage and had been ushered at the beginning of the nineteenth century into a scientific stage in which it limited itself to the observable. If one consults books on positivism, one will usually find that it is also characterised by a variety of additional features. Kolakowski[1] lists the following characteristics: phenomenalism, nominalism, rigid separation of values from knowledge, and the unity of science. Most of these features, I would prefer to think, are secondary and almost accidental. At the heart of the matter there stands the Olympian perspective stressed by Richard v. Mises in his classic *Kleines Lehrbuch des Positivismus* of 1939.[2] In Chapter 1 of his *History, Man and Reason,*[3] Maurice

Mandelbaum gives a slightly different description of positivism: first, it is a rejection of metaphysics; second, it assesses the adequacy of knowledge according to its approximation to the achievements of the most advanced sciences; and third, it confines science to a description of correlations. In a way Mandelbaum here begged the question, for to a positivist, the achievements of the most advanced sciences are the achievements which correspond to positivism.

If a closer definition of positivism is required, one should go to J. S. Mill's *The Positive Philosophy of Auguste Comte* of 1865, in which he stressed at the very beginning the phenomenalism of positivism; the rejection of the search for ultimate explanations; the insistence on the descriptions of regularities and the implied conviction that phenomenalism cannot mislead, provided we take its findings to be descriptions of appearances and not revelations of essences.

This model of knowledge comes indeed in many varieties. Each variety has its own difficulties. If one consults such an old-fashioned standard text as the Kirchner-Michaelis dictionary of philosophy first published in Berlin in 1886, one will be told that positivism was not just one of many philosophies; but the philosophy which stood behind the entire enterprise of science. In this dictionary we are told that positivism states that all knowledge is scientific knowledge because it is based on observation; and that there can be no other knowledge. The logical positivism of the Vienna Circle in the early part of the twentieth century was a bit more sophisticated; but not substantially different. The logical positivists believed that a subjective state such as 'observing' can be transcribed into a sentence, i.e. a protocol sentence. Carnap supposed that a *Protokollsatz* was a statement possessed of special epistemic security and free of theoretical presupposition. He did not consider that the impossibility of having a private language about observations might be an obstacle to such transcription. This formulation, euphorically, if unfelicitously, combined phenomenalism and empiricism and proved the Viennese true Comteans. Somehow they also relied too slavishly on Hume. Hume had maintained that sense-impressions were clear and strong; and ideas, vague and general. Unfortunately he had got it back to front. It is the sense-impressions which are vague and uncategorised and which remain silent in their travels around the circuits and maps of our nervous system. It is our ideas which are clear and strong, because they come from the realm of words and the rules which govern the use of words. The problem, never faced by positivists, is how these silent impacts come together with the not so silent labels. The coming together cannot be magical and certainly not – as positivists of all hues believed – automagical. The answer is simple enough: they come together hypothetically. But once the hypotheticality of this connection is admitted, the entire positivist programme turns out to have been a pipe-dream. For in this case, no matter how user-friendly or contact-friendly the relation between our body and the world is, such contact can never result in the production of sentences. Sentences are hypotheses and could be wrong – which dashes the positivistic hope to the ground. For that hope was that one could build up knowledge on the foundations unfailingly *caused* by this immediate and intimate contact.

Ideally, the positivistic programme was a conceivable programme. Why, after all, should one not be able to place the body in a suitable relation to the world and wait for the world to enter it and generate sentences? A language should be completely transparent and so constructed as to describe literally the impacts, no more and no less. But the more we learn from neuroscience, the less such a programme is conceivable. For neuroscience has shown that the impacts are silent and remain so and that a lot of hypothetical intervention is required to generate the worded representations of the way the world presents itself to the nervous system.

Positivism was, or appeared to be at that time, quite simple and straightforward and led eventually to the so-called Received View in the philosophy of science. The term 'Received View' was coined by Hilary Putnam in 1962.[4] 'Scientific theories', we are told by the View, 'are axiomatic calculi which are given partial observational interpretation by means of correspondence rules.'[5]

Though the term 'positivism' was not used by Locke, Locke makes a good starting point for an inquiry into this whole tradition. In surveying the rise and the decline of that tradition, one will find that, at the best of times, this tradition of positivism had very shaky foundations and that its rise was due to extraneous circumstances rather than to the viability of its own arguments. At its centre there stood an explicit theory about the acquisition of knowledge. 'I imagine that all knowledge', Locke wrote in the opening paragraph of Draft A of his *Essay*, 'is founded and ultimately derives its self from sense.' And in the very last edition of the *Essay* he wrote in Book IV, Ch. xi, section 1: 'No particular man can know the existence of any other being, but only when, by actually operating on him, it makes itself perceived by him.' We get notice of the existence of things by receiving instructions. Locke, in other words, believed that we can account for knowledge by thinking of it as sort of physical transfer from one thing to another. The world, in short, *causes* us to have knowledge of it. Locke's causal conception fits in very well with many modern conceptions of causation which think of causation as direct energy transfer and which deny that there is a causal action unless there is such energy transfer. 'Causation', Quine wrote in his *The Roots of Reference*,[6] 'is energy flow.'

Locke devoted a lot of attention to this problem of the causation of knowledge. He was aware that consciousness can interfere with the causal process because it is prone to 'fancying', to guessing and to believing.[7] Such imaginary activities have to be excluded and eliminated. When they are, the energy flow can proceed freely, and the ideas we acquire under these conditions are ideas of real things. Locke was aware that the causal generation of knowledge cannot proceed without the intrusion of ideas. For what the mind contemplates, he said, are ideas. It is ideas that are immediately present, not the things. Here we find an early combination of phenomenalism and empiricism.[8] 'How shall the mind when it perceives nothing but its own Ideas, know that they agree with things in themselves?'[9] His answer is straight: as long as consciousness does not interfere in the shape of willing, fancying, guessing, etc., we can take it that those ideas are caused by something in nature.

In spite of echoes of Bacon's *Idols*, Locke was nothing if not original. In all earlier attempts to account for knowledge, no matter what the precise nature of the mechanism of knowledge acquisition had been supposed to be, it had always been assumed that there is an established link between knower and known, so that the knower knows in advance, so to speak, what to expect, what to listen to and what instructions he or she is likely to receive. Needless to say, this link had never been conceived in terms of an evolutionary adaptation of the knower to his or her environment. But a link there was believed to be. For Plato, knowledge depended on ideas or forms which the soul had spotted before it was entombed in the body. Thus all human knowledge depended on the life-cycle of the soul which contained a period in which the soul, before its entombment in the body, had been part of the real world in which it could see the general forms of all the many objects it would eventually encounter during the term of its entombment. In St Thomas' Aristotelianism, knowledge depended to a large extent on the exercise of right reason. Both 'right reason' and the world to be known had been made by God and there existed, therefore, an ultimate link between knower and known, for both were the creations of God. For Descartes, to mention another example, knowledge proceeded from clear ideas which admitted of no doubt. True, here there seemed to be no initial link between clarity and what is known. But it so happens that for Descartes the primary clear idea from which everything followed was *cogito, ergo sum*: i.e. the awareness of a link between knower and known. For the famous formula asserts not only that I am *because* I think; but also that, in order to think, I have to *be* here. By contrast, Locke started a singularly novel line of thought. He not only argued that there is no link between the mind and the world; but alleged that the very fact that there is no link made the contents of the mind (conceived as a sort of bucket) a genuine guide to the world. Knower and known are distinct and without link. And when we finally *can* discern a link, he said, we can be sure that the world has exerted a causal influence upon the knower, and when that causal link has been detected, we can speak of having knowledge. The whole theory depended on the initial absence of a link.

We know today that this account of knowledge is completely false. To quote Donald Campbell: 'Somewhere in the evolutionary hierarchy the available distal relationships come to be exploited, and with this comes a renunciation of rigid one-to-one reflexes at the proximal level.' Proximal particulars (something like Locke's directly derived ideas) are equivocal, Campbell says. This equivocalness is corrected by a continuing process of pattern matching in which, for example, ideas in our mind are compared to the pattern of appearances as observed and gradually brought into line with each other.[10] However, we can see that Locke, though wrong, was not completely at sea and that one could correct his theory if one took evolution into account. Our sense-perceptions are adaptations to the environment in which we are living, and while we cannot get ourselves to believe that our knowledge of any part of that environment is literally caused by those parts, we can see that the adaptiveness of our senses provides us with an opening into the world. Yet Locke persisted in his theory without any inkling of the evolutionary amendments one ought to make to his view of perception. This is hardly surprising. In the seventeenth century, ideas of evolution were not even in the air, and Locke himself took great pains in the opening sentence of Book I of his *Essay* to stress that man is unique and totally different from all animals, i.e. to start from a totally non-evolutionary standpoint. If one were to amend Locke in the light of evolution, one would certainly be offending against the spirit and intention of Locke. And yet, there were people in the seventeenth century and earlier who had noticed, and Locke might have been one of them, that there was a certain adaptiveness between any animal and its surroundings. The explanations provided, however, were anti-evolutionary, not evolutionary. God in his goodness, it was argued, seeing that camels liked heat and sun and sand and little water, had taken care to place them in the desert![11]

Though the philosophy of observation would have been off to a flying start with the help of evolution, one can hardly blame Locke for failing to take advantage of the theory of evolution. But even so and even without evolution, had Locke been less concerned with the uniqueness of man, he would have seen that men, like animals, do not learn entirely by observation and by receiving instructions from the outside world. In man the complexity of the manner in which the nervous system meshes with stimuli may not have been obvious. An observation of animals would have helped. But here, evolution or not, there was a real barricade of dogma in the seventeenth century. No matter what observations even men like Bacon had made, people were thrashing around for reasons to make sure that man kept on being seen as unique. Nakedness, it was alleged, is bestial. Men wear clothes. When somebody mentioned that animals are not naked either but have hair and fur, people did not allow this observation to brighten their ignorance. They retorted instead that, in order to keep themselves unique, men had to cut off their hair because being covered by hair was a sign of bestiality.[12] And yet, men more enlightened than Bacon or Locke started having their doubts. Lord Monboddo thought that animals might learn to be human if they could live longer and Pepys in 1661 wrote of a strange creature from Guinea, an ape so much like a man that one had to think of him as a hybrid between a man and a baboon. Sir Ashton Lever (1729–88) had an orang-utan who had learnt to speak a few words.[13] In spite of such observations, the idea that man was unique and that his humanity consisted in his absolute difference from animals held firm for another two hundred years and left its fateful imprint on the rise of positivism. And yet, as is well known, Locke showed fewer scruples in regard to other pieces of Christian dogma.

Or consider Locke's theory in relation to the greatest success story of his age, Newton's *Principia*. Locke admired Newton enormously. 'Mr Newton in his never enough to be admired book', he wrote in Book IV, Ch. vii, section 3, and stated that Newton for one never made use of a priori 'maxims'. Newton himself seems to have imagined that he had followed Locke's theory as to how one gains knowledge. First, he said, one reveals by analysis some simple results and then generalises them by induction. At least this is how I. B. Cohen[14] summarises the scenario which Newton claimed to have followed. Going further into the matter, however, Cohen tells us that what Newton really did was totally different from this Lockian scenario. Newton's style, as Cohen calls it, is displayed in an alternation of two phases or stages of investigation:

> In the first, the consequences of an imaginative construct are determined by applying mathematical techniques to the initial conditions concerning mathematical entities in a mathematical domain. In the second phase the physical counterpart of the initial conditions or of the consequences are compared and contrasted with observations of nature ... This usually gives rise to some alteration of the conditions of the initial construct, producing a new phase one, followed by a new phase two, and so on.[15]

It takes very little discernment to see that in Cohen's view Newton did not follow Locke but, if one might say so, Donald Campbell. For that matter, if further proof is needed, the striking conceptual, and hence methodological, differences between the *Opticks* and the *Principia* show that in at least one of these works, Newton could not have followed Locke.[16]

The consequences of Locke's positivism and of his causal theory, devoid of an evolutionary corrective as they were, proved completely counterproductive. In Hume, they led to real scepticism and in Berkeley to a denial of the existence of the real world when it was not actually perceived. Hume thought that Newton had been a good practitioner of Locke and had proceeded by 'admitting no principles but such as were founded on experiment ... but [had been] resolute to adopt every such principle'. Hume's eulogy of Newton in chapter LXXI of *The History of England* is well known and often quoted. But Hume's ironical scepticism about Newton in the same paragraph is not so frequently quoted: 'While Newton seemed to draw off the veil from some of the mysteries of nature, he showed at the same time the imperfections of the mechanical philosophy; and thereby restored her ultimate secrets to the obscurity in which they ever did and ever will remain.' I take this to mean that Hume, certain that Newton had practised what Locke had preached, thought that Locke's philosophy had very serious limitations. Our own assessment of Newton is quite different from Hume's. Thanks to Cohen we know that Newton did not follow Locke and that, for this reason, his discoveries, though they have important limitations, are not nearly as 'obscure' as Hume thought.

Given the fact that Locke's prescriptions for and explanations of knowledge are not nearly as marvellous as he and Newton and Hume thought, how are we to explain that Lockian positivism, or something very similar to it, came to be one of the dominant philosophies of knowledge for over two centuries? I would like to put forward the theory that Locke's success and enormous influence is not to be explained by the inherent plausibility of his philosophical reasoning, let alone by the successes of their application; but by a peculiar view of human history and of the development of mankind. Locke simply reasoned, though he never spelt this out in so many words, that exclusive reliance on sense-perception is the last and most modern method for getting knowledge. This way we learn by taking instruction from the world. Every other method had been tried. People had believed in revelation and in tradition. They had paid heed to authority and had relied on innate ideas. Whenever they had done so, they had not achieved much knowledge. The whole past history of mankind, he knew, is full of superstition and ignorance, and he ascribed such superstition and ignorance to the reliance on faulty methods of getting knowledge. In his own age, by contrast, knowledge had made real progress. Since everything had been tried and had been found wanting, there was only one other method left. That method, he thought, was the causal theory of perception. It did not and could not have occurred to him that the reason why, say, revelation had never led to much real knowledge was that it had been shielded from criticism because it had been used as a social cement to keep society together. For that matter, if sense-perception and instruction were similarly shielded from criticism, they would not lead to much knowledge either. The fault had been the absence of criticism; not the peculiarly religious nature of the source of knowledge.

Ultimately, then, the real motive power in Locke was not philosophical reasoning, though he tried hard. The real motive power for his view of knowledge was the conviction that in the past men had tried faulty methods and that in his own age, men were at long last capable of using the correct method. The causal theory of perception was right, he thought, because it is the hallmark of the modern enlightenment. There is a sort of developmental law of history, he would have argued had he been pressed, which makes mankind advance from wrong methods for gaining knowledge to right methods. The causal theory of perception was right in virtue of a purely eliminative process. It was the one method which had never been tried before, and since its discovery coincided with the first genuine growth of knowledge, it had to be considered a valid account of that growth. Locke, in short, was something of a historicist: the main backing for this theory of perception did not come from philosophy at all, but from the historicist view that there was a law of development which had brought man finally to the stage at which exclusive reliance on causally produced knowledge, on knowledge as the causal interaction between knower and known, was the final stage of development.

In order to understand the full significance of Locke's non-philosophical motives, one has to look at the way in which he saw his position in the history of thought. Locke saw himself as a sort of revolutionary intellectual Hercules who had cleared out the Augean Stables. There are three pieces of explicit evidence to this effect. In the beginning, Locke wrote, all people were like the people who were then living in America, i.e. savages.[17] We have become enlightened and civilised; and they have not. Locke started his Treatises on Government by devoting the whole of the first Treatise to the demolition of a traditional view of the origin and character of government, i.e. the views of Robert Filmer. Filmer was no great intellect; but if one were to make a comparison between Locke's own theory and that of Filmer, one would have to admit that Filmer's theory, though illiberal, was much closer to the historical truth than Locke's. Finally, Locke's *Essay Concerning Human Understanding* begins with a lengthy attack on the theory of innate ideas. In all respects, one can see that Locke had no mean view of his own role in the history of mankind and in particular of his contribution to human enlightenment. Had Filmer and the theory of innate ideas been allowed to prevail, he seemed to be saying, there would be little difference between savage America and civilised England!

This assessment of himself was enthusiastically and unreservedly supported by no less a person than Voltaire. Writing to Walpole he called Locke 'another Hercules', and in his *The Age of Louis XIV* he uses the example of Locke to prove that the seventeenth century was superior to the golden age of Greece.[18] Such an evaluation of the significance of Locke in human history is all the more remarkable because it is not borne out by seventeenth-century historians like Sprat and Glanville, who saw progress in knowledge as very gradual and continuous. It had started with the Greeks and worked up to its culmination in the Royal Society, whose hallmark was experimentation. Strange as it may seem, Locke's view of himself as some kind of sudden fulguration was more in line with the thought of the Fifth Monarchy Men and other Puritan fanatics of the seventeenth century who believed that the millennium, intellectual or other, was at long last at hand. Here we can notice how deeply embedded the sudden illumination of positivism was in the darkest traditions of Christian phantasy. In his view of history and the suddenness of enlightenment, Locke seems to have belonged with the real obscurantists; and not with Sprat and Glanville! However this may be, we must concentrate on Locke's view of his position in history in order to understand how he could be so convinced of the liberating effect of his philosophy of observation and of the mind as a sort of bucket (to use Karl Popper's phrase) into which the world was flowing in order to produce knowledge. Although philosophically Locke's theory could not stand on its own two feet, it found great support in his view of history and of his place in that history.

Two centuries before Locke, at the time of the Renaissance, scholars and scientists had often looked back upon the past. But when they were looking back, they had only seen the Greeks and the Romans, to whom they felt, on the whole, inferior. They had therefore formed the idea that knowledge and civilisation was what the Greeks and Romans had achieved and that the future of mankind must depend on attempts to recover the knowledge of the Greeks and Romans. This peculiar, though understandable, vision of the past came to be punctured and eventually discarded by the awareness that with Galileo and Kepler and Copernicus knowledge had outstripped the Greeks and the Romans and that what came to be known as the Enlightenment had to be explained in some other way. How had the Enlightenment been possible? What was its place in the history of mankind?

One of the first men to address himself specifically to this question was Voltaire. The answer he came up with was very strange and, to us, totally unconvincing. But to appreciate its significance we have to consider that in the sixteenth and seventeenth centuries men had begun to see the past in a new perspective. Where the men of the Renaissance had identified the past as the Greeks and the Romans, observers in the sixteenth and seventeenth centuries, not to mention the eighteenth century, had come across the inhabitants of Africa and America, whom they thought much less civilised than the Greeks and Romans. Hobbes wrote in his *Leviathan*[19] that the natural condition of man was a state of war of everybody against everybody, i.e. the state of the savage people in many places of America. Fontenelle[20] declared in his work on the *Origin of Oracles* of 1686 that in the first ages of the world, men must have been as ignorant and barbarous as Kaffirs, Lapps and Iroquois are today. When Locke coined his phrase 'in the beginning the whole world was America', he was not merely echoing Hobbes and Fontenelle but could have invoked the support of countless sixteenth- and seventeenth-century ethnographers and travellers who had come back from Africa with reports that the crudest, most savage and most barbarous conditions were prevailing on that continent.[21] With such information, people in the seventeenth century were beginning to tumble to the idea that the original condition of all mankind had been like the conditions that were prevailing in Africa and North America and that when one is looking at the past, one has to consider that the past, even for people living in Europe, had been full of people who were ignorant and superstitious. In the Renaissance perspective of the past there could be no room for progress; but in the modern perspective of the past, which was not confined to the Greeks and the Romans but took in Africa and America, there was ample room for progress. This progress, they reasoned, had at long last taken place.

When Voltaire began to consider the matter, he came up with a theory of history which is very similar to the theory of eighteenth-century geologists. They explained the mountains and oceans and valley and rock formations on earth as the result of gigantic, sudden, unpredictable and inexplicable catastrophes. Voltaire applied the same kind of reasoning to the Enlightenment. Looking back he saw that he was living in an age of enlightenment and admitted that there had been comparable ages in the past. He identified four such ages: the Athens of Pericles, the Rome of Augustus, the Florence of the Medici and the Age of Louis XIV. Unpredictably and catastrophically, we might say, cultural volcanic eruptions of enlightenment had pushed their way up and created temporary islands of civilisation and knowledge. As unpredictably as they had arisen, they had subsided again. He was sometimes a little bit more optimistic about the future of the enlightened age in which he himself was living and hoped that the star of Newton had come to stay.

To historians such a catastrophist theory of the Enlightenment was not satisfactory. It left more questions open than it was able to answer. Why the sudden eruptions? How were they related to surrounding periods of history? Why had they subsided? And last but not least, what was to be the future of the present 'catastrophe' which had been presided over by Locke and Newton?

Voltaire's 'catastrophist' explanation of the liberating effect of Locke and Newton was inspired by the reasoning which had presided over the famous quarrel between the ancients and the moderns. This quarrel had broken out when the first doubts had arisen in the seventeenth century about the unsurpassed excellence of the Greeks and Romans. Some had argued that that excellence was final, even when compared with the achievements of Galileo and Harvey and Newton; and others had argued that the achievements proved that that excellence had not been final. Either way, the argument was catastrophist in that it had concentrated on the relative merits of two catastrophes, one ancient and one modern. The participants in the debate were not arguing about history or comparing competing views of history but kept arguing about the relative merits of two different catastrophes.

Though Voltaire's catastrophist explanation of the Enlightenment was by far the most readable work of the historical literature on the subject and was to have far-reaching effects on the study of social and cultural history in the nineteenth century, Voltaire remained alone. During the eighteenth century, there came a whole crop of more genuinely historical explanations. These explanations employed a very different model of history. They worked on the assumption that there was a law of development which had propelled mankind from an original state of savagery and superstition to its present state of knowledge and civilisation. In its most specific form this law of development stated that society had grown naturally and progressed normally over time through four, more or less distinct, consecutive phases. The stages were originally defined in terms of different modes of subsistence: hunting, pasturage, agriculture and commerce. It requires little exercise of the imagination to see that once the law of the four stages was fixed, in terms of the mode of subsistence, one could easily elaborate it and assign to each mode its prevalent type of ignorance. Starting with Locke himself, who did not elaborate the law, we can trace its growth and hold in Bossuet and Fontenelle and Turgot right down to the end of the eighteenth century, where it dominated the great work on the history of America by William Robertson, who actually spelt out that once the mode of subsistence in every stage is known, one can find out how it affected, in every stage, laws and policy. The law of the four stages was a developmental law and, as such, historicist. It asserted that there was an iron necessity which compelled progress through these four stages and that the sequence was a unique succession of events. There was no need for further inquiry.

One must ask, however, where Turgot and his fellow-historians got this idea from. Ronald Meek[22] points out that the idea that there had been different stages in the history of mankind was a very old idea and can be traced back to the ancients. He says it had been dormant for two millennia when it was eventually taken up in the seventeenth century by Grotius and Pufendorf. The ancient writers, however, had been in two minds about the direction of the sequence of these stages. Hesiod had seen a decline and deterioration from a golden age to an iron age and others, from Plato to Machiavelli, had seen merely repetitive cycles.[23] It was only from Bodin and Bacon onwards that the idea that the sequence of stages was taking place in an upward direction and linked to progress through enlightenment gained a firm foothold. This notion of progress came to be very firmly established by the early eighteenth century and is apparent even in so theology-minded a work as Bossuet's *Discourse of Universal History* of 1681, which was otherwise completely out of step with the Enlightenment and a target for both Voltaire and Turgot.[24]

The derivation established, we are still left with the inherently unreasonable character of this law of the four stages. It was unreasonable because it could not explain why there had been a singular succession of those stages and why that succession should be a developmental law which explained what had happened and why modern Europeans were not like the inhabitants of Africa or America in the seventeenth and eighteenth centuries. All the same, compared with the traditional theology-dominated histories which had been extrapolated from the Bible, the historicism of Turgot was yet another mark of enlightenment. It seemed indeed more enlightened to assert that there were four stages in the development of mankind than to say that God had providentially guided the evolution of mankind from the Garden of Eden to the liberalism of the commercial culture of eighteenth-century Britain and

France. Whatever the faults of historicism, it was a net gain over history inspired by theology.

The net gain did not just consist in the fact that it was non-theological and did not invoke God and providence and, therefore, had a certain explanatory power. It also looked 'scientific' because if one makes the assumption that there are given wholes, phases, styles, stages or systems, one can make the *observation* that the sequence of these wholes reveals laws of development. This method of arriving at such laws of development is not really scientific, because it is entirely based on the assumption that there are such wholes. But it looks scientific, and is certainly more scientific than the theological view which had preceded it, i.e. that there are divinely ordained sequences . A. F. v. Hayek has therefore, rightly, called this kind of history 'scientistic' history. And this is what we mean by 'historicism'.[25]

The historicist law of the four stages was also an improvement on the simple Renaissance view that excellence had been present in Greece and Rome and had to be recovered in the fourteenth century. In this view the middle ages were simply the 'age in between' the two great cultural peaks of mankind. With the law of the four stages this view of the middle ages changed. The middle ages ceased to be a period of Gothic depravity. Turgot saw the middle ages instead as the stage of agriculture and therefore as an improvement on hunting and pastoralism. Eventually Comte adopted an even more positive attitude to the middle ages. He recognised that the democracy which had prevailed in the personnel of the church had helped to lay the foundations for a more liberal order of society and had detracted even then from the authority of magic and religion which had been linked to a hierarchical order of society or a patriarchal order which had sustained magic and religion.

The historicist law of the four stages was also a great improvement on the catastrophist theory of Voltaire, even though according to our own understanding it explained nothing at all. What, one might ask, should have compelled this particular sequence? Why was there compulsion at all? Could it have been the other way round? There was, however, to be no improvement on this kind of historicism until the model of constructing historical sequences of Lyell came to be known. Lyell eventually showed that one can explain past events by the operation of the causes which are acting in the present.[26] With this methodology he drew away from historicism and showed the way to a proper understanding of historical development. In the mean time, however, historicism was accepted and served as an explanation of the Enlightenment. Such an explanation was indeed necessary, for, as we have seen, the philosophical reasoning which had recommended sense-observation as the royal road to success was extremely wanting as a theory and was not practised by those men who had actually advanced knowledge and brought about the Enlightenment. Without the historicist explanation that sense-observation was 'enlightened' because it was the philosophy of the fourth stage of development, the philosophy of sense-observation or any other form of positivism could not have won support. With historicism, it was firmly established in the face of arguments and evidence to the contrary. Thus there began one of the unholiest alliances in the history of thought: the alliance between positivism and historicism. Positivists paraded their philosophy on the grounds that it pertained to the most recent stage of development; and historicists paraded their developmental law on the grounds of positivism. A careful and disinterested study of history based on the rational criteria required by positivism – absence of authority and of hearsay, avoidance of wishful thinking and of mere legend – had led to the discovery of the developmental law of the four stages. It took a very long time, incidentally, to remove this kind of positivism from historical study. It dominated the historical researches of Karl Marx and Leopold von Ranke and showed up in the very title of Arnold Toynbee's great work *A Study of History*. And through Ranke, it came to be academic orthodoxy in our universities and presided over the establishment of the first chair of modern history in the University of Oxford in 1850 when the University authorities, seeking a substitute for the theology-dominated curriculum which had led to nothing but controversy, hit upon the study of history as the most uncontroversial study because it was the most positivistic kind of study available at that time.[27]

It was believed that the study of history was based on documents and that, provided one did not stray from the documents, one could not make mistakes. Such documentary positivism as the method of historical research was the precise parallel to Locke's positivism of perception. The process by which one gains historical knowledge through documents was a little more complicated because one cannot describe the relation between the historian and the documents in the causal terms in which Locke had described the relation between man and nature. But historians tried hard: source criticism was the answer intended to nullify bias and thus assimilate the relation between historian and sources to a causal relation. In an impassioned moment, the great French historian Fustel de Coulanges leant towards his students and stated: 'It is not I who am speaking, but history itself.' And if there are doubts about such positivism in historical knowledge, I will quote a passage in which Richard v. Mises.[28] summed up positivism in *historical* knowledge: 'The criterion of truth of an historical assertion lies in the testability of the still observable aftereffects of the alleged fact and of its indirect consequences (sources).' Less sophisticated historians (and practising historians often are very unsophisticated) cut this process of testability short. They imagined that in writing history they were simply copying from History, i.e. they were transcribing events into words. They thus imagined the causal process to take place between what happened and what was being written so that historical narratives were 'caused' by events. R. v. Mises is a little more circumspect and realises that the causal relation is between the *after-effects*, among which he includes the sources and documents, and the composition of the narrative.

With views like these there was established the positive science of history. History itself came to be seen as part of positive knowledge. Positivists, it is to be admitted, looked askance at historians like Hegel and Spengler in whom theorising had outstripped the study of the sources. But barring such extremes, historical knowledge was enrolled among the positive sciences and historical knowledge positively told of the continuing progress of mankind from savagery to enlightenment. This had been the burden of Turgot's story. It was explained by Herder in his *Ideas on the Philosophy of the History of Mankind* and found its greatest exponent in H. T. Buckle's widely acclaimed *History of Civilisation in England*, which Mises called 'a model of positivistic writing of history'.[29] Buckle, he seems to have meant, used the positivistic method for writing history and thus produced a history which explained that positivism was the correct philosophy of knowledge because it provided historical evidence that theology and metaphysics had been superseded by the positive reign of reason.[30]

I have dwelt on this alliance between positivism and historicism at some length because it is crucial. Positivism by itself was poor philosophy and could not have gained much of a hearing. Thanks to Karl Popper we know that historicism by itself was poverty stricken.[31] But when positivism could invoke a historicist theory to say that the reign of positivism had to come; and when historicism could invoke the method of positivism and *look* like science to support the contention of positivism, these two philosophies became indispensable to each other.

The alliance between positivism and historicism found its grandest exposition in the work of Auguste Comte in the early nineteenth century. Comte, moving away from subsistence as the definition of each stage, fastened upon the mode of knowledge as the characteristic which presided at every stage. Thus he reduced the stages from four to three and taught that mankind had progressed from magic to religion and from religion to positivism, by which he meant modern science. If one peruses the voluminous writings of Comte, who did more than any other philosopher to give the word 'positivism' its status and currency, one will find next to no philosophical argument in favour of it. When he did try to provide a philosophical argument, it led from muddle to muddle. 'The question to which it is exceedingly difficult to find an answer in Comte's work', A. F. v. Hayek writes,[32] 'is what precisely is meant by the "phenomena" which are all subject to invariable laws, or what he regards as "facts".' Comte, instead of dispelling such and similar doubts, based his advocacy of positivism entirely upon the fact that it was the mode of knowledge of the third and final stage of the development of mankind. We can see the interdependence of history and philosophy, i.e. of historicism and positivism, very clearly in Comte. In the scientific stage of knowledge, when metaphysics and theology have been superseded, scientists rule out questions about the hidden nature of things. They cease to ask: why? and ask instead: how? This was the philosophy of positivism. How could it be recommended or legitimised? Comte legitimised it by an appeal to history. The iron law of development decrees, he argued, that theology and metaphysics are superseded by science and in the last stage of development, when enlightenment has arrived, the intellectual world is dominated by scientists, i.e. by people who know the relation between the external world and man, and this knowledge enables them to present a 'system of positive knowledge' in which every hypothesis must be capable of positive verification.[33] This was the historicist view of history. How could it be justified? By the contention that a positivistic treatment of the sources would reveal that the course of history was from theology to metaphysics and from metaphysics to science. Rarely in the history of thought has there been a finer example of how thoughtful and highly respected people were pulling themselves up by their own bootstraps.

If we are to believe John Stuart Mill, positivism's view of science 'became the general property of the age'.[34] The last generations of the nineteenth century were 'as if mesmerised by a very rigid conception, a truly Comtian conception of natural science … Such was the nearly unanimous opinion at the time.'[35] But we have to be careful. The great purveyors and advocates of positivism from Bacon to Locke and from Comte to Mill, Mach and Carnap, with the one important exception of Mach, were not practising scientists themselves, and the methods which they recommended to scientists were not 'always necessarily those', F. A. v. Hayek writes,[36] 'which the scientists in fact followed in their own field, but rather those which they believed they employed'. We have already noticed the gap between Newton's professed method and his actual style. We will find similar doubts and discrepancies between profession and style in the nineteenth century. Maxwell frequently expressed himself as a positivist,[37] but was also capable of a different frame of mind, as is apparent from his essay "Molecules" of 1873, where he sought to derive a proof for the existence of God from an examination of molecular structure.[38] God or no God, with or without molecules, it is impossible to imagine that his famous equations could have been inductively derived from observations. It is much more likely that he employed a style of investigation similar to the one Cohen says Newton employed. Though Faraday was never at his best with pure theory, he struggled violently against the demands of positivism because, as is well known, 'lines of force' are not easily susceptible to the requirements of positivism.[39]

If we consult L. Pearce Williams' biography of Faraday,[40] we do not get the impression of a man too preoccupied with the philosophy of his pursuits; but Williams described Faraday's philosophical education in some detail in his second chapter. He seems to have been anything but a positivist. For instance, in 1812, 'when he first made Davy's acquaintance, he was a confessed and ardent believer in the real existence of imponderable fluids. Davy, in the first lecture Faraday attended, soon shook this belief'.[41] Davy was no positivist himself and had been close to Coleridge, who had introduced Kant into England. 'The result was that Faraday found himself in an intellectual muddle', Williams concludes.[42] Helmholtz, to cite another example, wrote that 'the appointed task of physics is thus to refer natural phenomena to unchangeable attractive and repulsive forces, whose intensity depends upon distance'.[43] With this view one can hardly call him a positivist.

Kirchhoff seems betwixt and between. He declared that the mechanist view could not explain phenomena – only describe them. This was a good piece of positivist thinking. But when he continued that his goal was the complete description of the motions occurring in nature and to exclude other phenomena from such description, he seems to have gone beyond the limits set by positivism, at least in the sense in which Mach understood them.[44] With Hertz one remains equally unsure. He argued that no objective content need be sought in the system of Maxwell's differential equations over and above that already expressed there. We can, he continued, deduce consequences from Maxwell's theory and find them verified by experiment. This is all the proof of the theory's validity we can get, and no more is possible.[45] 'Experience', he wrote,[46] 'remains wholly foreign to the considerations' of time, space and mass. 'They rest on the laws of an inner intuition and forms of the inner logic of the person stating them and have no connection with his external experience other than these intuitions and forms may have with it.' This way of thinking was Kantian and certainly not part of positivism. And Poincaré, as is well known, was not a positivist but a conventionalist.

The case of Darwin is specially interesting. *The Origin of Species*, i.e. the way he set out his theory, is often described as a classic of inductive logic and could thus be considered a work informed by positivism.[47] However, we know from Darwin's notebooks as well as from his letter to Haeckel of 8 October 1864 that his method of discovery owed nothing to positivism. In his observations on his journey on the *Beagle* he was selective, and it was this selectivity which made him, in October 1838, so receptive to Malthus.[48] 'I happened to read for amusement "Malthus on Population" and being well prepared to appreciate the struggle for existence ... it at once struck me that under these circumstances ... '.[49] True positivists do not read for 'amusement', and they are not suddenly 'struck' by theories but compile them inductively and systematically by heaping observation on observation. Darwin was under pressure from fellow-scientists who believed him to have proceeded recklessly and thought that his theory might therefore be flawed;[50] and it may be for this reason that towards the end of his life he maintained, rather incorrectly and in contradiction to his own testimony, that 'my mind seems to have become a kind of machine for grinding general laws out of a large collection of facts'.[51]

Planck was not exactly a positivist either.[52] Boltzmann seems to have wavered at times.[53] Einstein, although at the very beginning of his career he had, as Heisenberg reminded him in a famous conversation in Berlin in 1926, sometimes flirted with positivism of the Machian kind, rejected positivism in a most decisive manner and insisted that only theories can decide what one is capable of observing and that observations by themselves are neither here nor there. When Heisenberg reminded him that in the Special Theory of Relativity, "simultaneity" had been defined entirely in operational or observational terms, Einstein replied:'Yes, that may be true, but it is nonsense all the same!'[54] This is not an exhaustive list, but these examples show that positivism did not exactly enjoy wide support among practising scientists, i.e. among people who provided the knowledge we have. Given the dubious credentials of positivism *without* historicism, this is hardly surprising. The only thing that is surprising is that in spite of so much evidence to the contrary, positivism came to be given a new lease of life by the Vienna Circle during the early part of the twentieth century. The reason for this new lease of life was neither the philosophical cogency of the argument that good knowledge has to be based on observation and on nothing else nor the actual practice of leading scientists, but the historicist conviction that observation-based knowledge was the most up-to-date and advanced kind of knowledge and must, therefore, be the most commendable kind of knowledge. The logical positivism of the Vienna Circle relied heavily on the emancipation and enlightenment historicism of earlier positivists and displayed a well-nigh political enthusiasm for the intellectual and moral liberation of mankind which they believed would be promoted by their view that good science was based on positivism.[55] In reiterating that science, being 'positive', is the knowledge-producing enterprise *par excellence*,[56] the Vienna Circle was on very shaky ground, for one glance at the practice of nineteenth-century physicists would have sufficed to prove the contrary. Science, to be sure, has produced a lot of knowledge; but not necessarily because it was 'positive'.

Nineteenth-century social scientists, on the other hand, were much more explicit and avowed both their positivism and their historicism. They failed to diagnose historicism as scientistic and took it to be scientific. It is in the thought of men like Maine, Tylor, Marx and Frazer that we can recognise the link between positivism and historicism best.

Frazer was a Comtean historicist. He believed that there was a developmental law which had determined man's intellectual evolution from magic to religion and from religion to science. When questioned about the reasons for believing that this law was true, he replied that a study of history taught that it was. In other words, he was a positivist who believed that simply by looking at history, one could determine the law it was governed or directed by. His historicism was founded upon positivism. Had one taken it the other way round and asked him why he was a positivist, he would have replied that positivism, which freed one from the superstitions of religion and magic to which men in earlier ages had been captive, was the philosophy and methodology of every intelligent man living in the nineteenth century and that the course of history had been directed towards the evolution of positivism. His positivism, in turn, was founded upon his historicism.

The nineteenth-century alliance between historicism and positivism has been obscured by a careless way of looking at the history of ideas, recently reinforced by the writings of Michel Foucault. Both Frazer and Darwin and countless other people used the term 'evolution'. Many books have been written about the use of the idea of evolution in the nineteenth century, and in all those books Frazer and Tylor, Comte and Darwin and Wallace and Lyell are seen as exponents of 'evolution', be it of the earth, society or species. There was, however, a world of difference between the uniformitarian evolutionism of Lyell and Darwin[57] and the historicist evolutionism of Comte and Frazer. Lyell and Darwin saw evolution as the result of the operations of general laws.[58] Comte and Frazer believed that there was a law of evolution which had determined the progress of mankind. The authors of books on the intellectual history of the nineteenth century rarely make this distinction and therefore find it very difficult to explain why the evolutionism of Darwin has survived and shown progressive problem shifts; while the evolutionism of Comte and Frazer is dead, though not necessarily because it showed degenerative problem shifts.

Michel Foucault has now increased this confusion by the peculiar theory of cognitive relativism he professes. According to Foucault the intellectual climate changes approximately once a century. Since he does not relate these changes to anything at all, one can only presume that he believes in some kind of number magic in which the figure '100' casts an intellectual spell over intelligent people. However this may be, Foucault diagnoses that in the nineteenth century intelligent people were given to explaining everything in terms of evolution. Thus he throws Darwin and Frazer together and obliterates the fact that Darwin's theory of evolution was uniformitarian and Frazer's historicist; and that historicism was based on positivism and that Darwin's evolutionism was not. Broadly speaking, Foucault can make a *prima facie* case – for the idea of evolution was very pervasive in the nineteenth century. But this very fact shows how important it is to make the right distinctions and not to throw things together that merely look alike. If Foucault's method were to be adopted across the board, one would end up by classifying flies with aeroplanes. No doubt, they have similarities. But are these similarities significant or accidental?

Leaving Foucault aside, let us now turn to the last great exponent of positivism in science, Ernst Mach. Mach was a towering figure and exercised a great influence on some of the finest minds of the twentieth century. Mach, unlike most other scientists, was an arch-positivist. He firmly believed that only those statements are true which are directly produced by sense-observation. In this way he came to question the existence of Newton's absolute space as well as the existence of atoms. In the first case he was working in the right direction, though for the wrong reasons; and in the second case, he was on the wrong track altogether.

However this may be, I have not been able to find any clear indication either in John Blackmore's book or any other work on Mach I have consulted why Mach believed that only sense-observation afforded access to the truth. Given the fact that this doctrine of the absolute veracity of sense-observation is so peculiar and given the fact that it has had such a chequered career ever since Locke, it is really very astonishing that a man of Mach's intellectual power should have placed so much reliance on it.

The only conceivable answer is that Mach simply took it for granted that exclusive reliance on what our senses tell us is more 'enlightened' than deference to any other source of information. There can, indeed, be no other explanation. By any critical standard, if one has sense-observations one has knowledge about one's self. When my skin is stung or my eyes stimulated, I can be fairly certain that there are events on my skin and on my retina. *Sensa*, so called, no matter how one twists and turns, are *sensa*, i.e. events in our nervous system and, possibly, in our mind. One has to make an act of faith in order to persuade oneself that they represent, let alone correspond to, something else, e.g. something that is happening outside one's nervous system. I cannot believe that Mach was the sort of man who would make such an act of faith. We are thus left with the only possible explanation. Like all other positivists, Mach was a historicist. He believed that sense-observations are the most advanced and enlightened form of information. The developmental law had forced Mach's ancestors to jettison first magic and then religion, and had left him with the truly enlightened and rational method of getting knowledge – reliance on sense-observations. Mach is yet another striking example of the contention that without historicism, positivism is untenable.

Philosophically, Mach's protestation that basic sentences are denotations of direct sensations such as 'I see blue'[59] is not different from Locke's causal theory of perception. In the two centuries between Locke and Mach, to the best of my knowledge, nobody had ever come up with a sensible argument in favour of this kind of positivism, though it had been formulated in many slightly different ways. One way or another, everybody had fallen back upon historicism: positivism, so the ultimate argument went, is more enlightened than theology and metaphysics. Mach was no exception, and Richard v. Mises, a positivist himself, says that Mach was the second stage in the 'emancipation and humanisation of knowledge'.[60] Mach insisted, and thus he emancipated knowledge, that correct knowledge is an extension of man's practical activity in man's relation with the world. Neither religion nor theology nor metaphysics was so related or able to be seen as an extension of man's practical activity. The way Mach fell back upon historicism is neatly illustrated by the Introduction he wrote to the German translation[61] of J. B. Stallo's *The Concepts and Theories of Modern Physics* of 1881. Mach[62] praised

Stallo because he had recognised and definitely removed the scholastic and metaphysical elements that still clung to the older physics.

The history of the unholy alliance between positivism and historicism had an ironical epilogue in the work of Wittgenstein. Whatever one thinks of the *Tractatus*, one must give it credit for being an attempt at positivism which did not invoke the idea that sense-observation was more enlightened than other sources of information. Wittgenstein made a genuine effort to show that there is a real and intelligible relationship between the sum total of all facts and the sum total of all true and meaningful sentences. This was precisely what all positivists had assumed. But like all other positivists, he too was not able to produce a rabbit out of a hat. His connection between the sum total of facts and the sum total of sentences was based on the quite non-positivistic assumption that there are what he called 'atomic facts'. Such an assumption flies in the face of what we have come to know about the relation between our nervous systems and the world it is in interaction with. The stimuli the organism receives do not result from 'atomic facts', but come as polymorphous sets. They are neither precategorised nor ordered. They are 'samples from polymorphous sets composed from objects and events in the world, the size and disjunctive partitions of which are limited by adaptive shaping during the natural selection of both sense organs and musculoskeletal systems'.[63] In view of my analysis of the rise of positivism it cannot come as a surprise that he had to base his positivism on a non-positivistic view. Other positivists before him had also had to rely on an auxiliary hypothesis, i.e. the hypothesis that there was a historicist developmental law which had made men enlightened enough to rely on observation as a source of knowledge. Wittgenstein simply exchanged the auxiliary historicist hypothesis for an auxiliary metaphysical one. For the rest, he proclaimed to have achieved what all positivists had hoped for. Positivism, it had always been alleged, supersedes all philosophical attempts to explain knowledge and makes philosophy superfluous. If one is a positivist, one knows that everything one knows is known scientifically by observation and that science thus takes the place not only of magic and religion but also of philosophy. Every reader of the Preface to the *Tractatus* knows that this is precisely what Wittgenstein claimed.

Let us return for a moment to the case of Mach. Unlike almost all other positivists, Mach had an appreciation of the relationship between our ability to take in the world through our senses and the fact that the senses belong to bodies which must act in the world. Although he did not refer to evolution and natural selection, he appreciated that there must be a certain connection between the body's ability to act in the world and the body's ability to sense the world. There must be a connection between the way the world presents itself and the way it is re-presented as a result of these presentations. In this sense, Mach's phenomenalism was not arbitrary, but based on an appreciation of the fact that the phenomena we experience must be intimately related to the way the world is. The human body is part of the world: 'The material world rests upon established connexions between elements, and relations between human impressions are only particular instances of such connexions.'[64] As we shall see in Chapter 4, such positivism cannot succeed, even when it is based more directly and squarely than was done by Mach, on evolution and the adaptiveness by natural selection of the senses which register the phenomena. But, historically, it is of interest that there is one strand in positivism which foreshadows the philosophical Darwinism which will be developed in Chapter 4.

3

THE LURE OF SOCIOLOGY

I

Even if the arguments traced in the two preceding chapters had been more successful, we would be left with a sizeable problem. When we started from within, we managed to obtain either labelled states of mind or symbols and were able to reach the conclusion that the symbols amount to a picture of the world which represents our states of mind. This conclusion is the inverse of the old situation in which the states of mind were somehow allowed to represent the world. Labels, however, even at their best, we found to be only hypothetical. Symbols, we found, though more immediate in their meaning and less hypothetical than labels, did not stand any more than labels in an unequivocal relationship to the states of mind they mean. Either a symbol means too many different states of mind because it is lacking in precision; or a whole series of different but more precise symbols mean one specific state of mind. In symbolisation there is either lack of precision in the *relatum* or uncertainty in the choice of a referent. Either way, we had to conclude that states of mind can neither be labelled nor symbolised with certainty.

Without certainty of articulation, a state of mind, contrary to the expectation that our own state of mind is the one thing we are certain about and have privileged access to, is unstable and inauthentic. Autonomy of our states of mind is comparatively easy to achieve. All one needs for autonomy is a determination to say 'no' to every hypothetical label or to any symbol proposed. But authenticity requires a more positive conviction that any label or symbol is really and truly adequate. There are many different reasons for the difficulty of achieving authenticity. We know that people who are firm about their beliefs and convictions – who dogmatically and inflexibly cling to their labels and symbols – are people who are insensitive to the nature of their consciousness. For they mistake a belief which hypothetically labels or symbolises the inchoate feeling they are conscious of for the conscious feeling. The people who parade as the hard-headed, tough-minded realists are, in reality, soft-minded delusionists. Moreover, one of the commonest and most potent sources of both guilt and shame is the uncertainty we have about our states of mind. When we do something and then hypothetically label what we take to have been its cause, we cannot be sure that that hypothesis is correct. One often hears it said that a certain person is acting in 'bad faith' – meaning that he or she knows what he or she *really* means to do but is not doing it or is concealing it. But the problem of being in 'bad faith' is much more complex than a mere imputation of duplicity or concealment would suggest. The real problem of 'bad faith' is that the person concerned does not himself or herself know what precisely it is that he or she is concealing. There can, therefore, be no certain knowledge as to whether there is or is not 'bad faith'. We then wonder what other people's hypotheses about our state of mind might be. And finally we get anxious, because we cannot make up our mind which of these conflicting hypotheses to reject. The situation is then further aggravated because we cannot really know what other people's hypotheses about us really are; we can only surmise. And all we really know is that our own hypothesis is not necessarily right. The failure to achieve authenticity does not just create a personal problem for the person who is failing in authenticity. He or she is very likely to become suspicious and even paranoid as to what other people's hypotheses are, and such suspicion and paranoia are further reinforced by the natural inability to be certain about one's own hypothesis.

Nietzsche was one of the first thinkers to recognise the problem of the instability of self-identification. The man who 'runs away from himself', he wrote,[1] 'hates himself, damages himself, is certainly not a good man. For he finds refuge from himself only in *other people*.' Nietzsche, however, believed that rushing towards other people was unnecessary and that the strong man *could* know by himself what and who he is. Sartre ended his famous play about hell with the observation 'hell is other people'. But he was only half right. Our hell of doubt and paranoia is not just because we are aware of the hypotheses other people inflict on us; it also comes, and comes mostly in the first place, from the doubt we ourselves must have about any hypothesis we might use about ourselves. One might even amend Sartre and say, on the contrary, that real hell is to be alone. Alone with one's hypothesis, one can spin it out. One may well doubt its adequacy; but when there is nobody to contradict and suggest an alternative, one frequently becomes captive to any one hypothesis one fancies. One spins it out, and then, by adding something one thinks one has heard, one interprets any event as proof that it is more than a hypothesis. As long as one is alone with oneself, there can be no limits to such flights of fancy. One can even become hardened in them until one actually begins to avoid the company of other people lest it pulls one back into the reality in which one realises that one is dealing with nothing more than tentative hypotheses. The old and trusted injunction 'know thyself' continues to sound good. But what precisely is there to be known over and above a bundle of hypotheses?

When we started from the front, we fared even worse. The information that came in got bushed and would not yield any kind of representation at all. Suppose, for argument's sake, we are prepared to make an act of faith and hold, as all true positivists of all hues have done, that the information, instead of remaining bushed, erupts, magically, in the shape of a sentence which then 'represents'. We would still be left with the problem of reference. How would such a magically produced sentence refer to the objects and events which are in front of us and which we believe to have caused the sentence?

Many philosophers have tried their hand at this problem of reference and have come up with all manner of explanations as to how such sentences hook into the world. It would be no exaggeration to say that theories of reference are the most arcane, esoteric and prestigious efforts any philosopher has ever been capable of. The virtuosity displayed in this field is directly proportional to the failure. For one can demonstrate that all these efforts must be condemned to failure. For a theory of reference to succeed, one would have to be able to state what the sentence in question is referring to without using that sentence. For only then could one analyse whether the reference-relation specified by one's theory of reference holds between the sentence and the object. Since it follows by definition that the object or event in question cannot be located *without* the sentence supposed to refer to it, no theory of reference can succeed; or, better, even if it did succeed, one could not show that it does. One can also put this in a slightly different way. If a sentence refers to an object and even if we somehow know what this object is, such knowledge would have to come again in the shape of a sentence. We are then left, time and time again, with a situation of reference in which one sentence refers to nothing more than another sentence. Any attempt to establish reference achieves nothing more than to bind us even more firmly and intricately into the web of language. The question as to what the original sentence and all the subsequent sentences *really* refer to remains wide open, because with one sentence after another, the actual object alleged to be referred to recedes further and further into the background, becomes more and more intangible and eventually dissolves into fog. This situation is analogous to an experience after a dream. When we wake up, the dream is vividly in front of us. But as soon as we start to seek words to express it, it recedes, becomes nebulous and finally vanishes altogether. All the technical expertise of the most dexterous thinkers has not been able to show that a sentence can reach *beyond* language. The last outcome of countless fine and technical distinctions is that all theorising about the relations between thoughts and languages on one side and objects on the other has to be done from within thought and language, and not even Gareth Evans[2] has succeeded in showing that the signified does not, in the end, slide under the signifier.

One would expect that both the problem of the correct identification of states of mind and the problem of what it is sentences refer to can be solved by a proper grasp of reality. If one could pin-point the state of mind and if one could pin-point the object or event intended or referred to by the sentence, both problems would be solved. However, it is precisely because we are unable to do such pin-pointing that these two problems cannot be solved. There is, however, a different kind of solution. One can shift one's search for a solution from looking at reality to looking at one's relations with other people. One can shift the problem from reality to society.

II

This shift, which has finally led to the belief that the conditions of sociability are the determinants of knowledge and that knowledge is not a relation between knower and known but a relation between several knowers, is one of the most striking developments in contemporary philosophy. The information which makes up knowledge is, so to speak, seen to come neither from within nor from the front, but from behind. Our knowledge, so the belief states, is not determined by the thing we know to have sensed or perceived. Nor is it determined by what reality instructs us to be the case. Our knowledge is determined by the communal relations we form with other people. Umberto Eco[3] gives a succinct summary, in post-modernist terminology, of this possibility. The meaning of a signifier, whether sign, index or icon, he states, does not depend on any reference to the real world. When we see a picture of a house, we do not recognise it as a house because it *is* a picture of a house and because it *refers* to a house, but because we establish a relationship between the picture and a unit of an internalised cultural system which we habitually call 'house'.[4] Seeking refuge from the instability of self-identification and the inscrutability of reference to the outside world in the authority of an epistemic community is a move which is ready-to-hand. People get themselves circumcised by baptism or some other initiation ceremony which commits them to a language-game or a life-form, so that their lives are rounded with a dogmatic sleep. Thus all worries about instability of identity and inscrutability of reference disappear below the horizon. And if political circumstances or socio-economic pressures tend to loosen the grip of the rules, they move towards the communal commission of violence which creates an instant society that reconfirms the labels they require to impale their vacillating states of mind on. Thus every conceivable text is seen to have a 'structure of meanings that is obvious and inescapable from the perspective of whatever interpretive assumptions happen to be in force'.[5] Recourse to evidence is to be precluded because it destabilises and causes uncertainty. There is, Derrida cogently keeps arguing, 'nothing outside the text'.

Michel Foucault managed to put it the other way round. As against the common belief that the Enlightenment, for example, broke social traditions and barriers in order to make recourse to evidence and reason possible, Foucault sought to persuade us that, on the contrary, the whole Enlightenment started a nightmare of ever-harsher social controls. Such social, as opposed to cognitive, controls were necessary to establish the purely arbitrary belief (mere discourse in his eyes!), so valuable to the power of the middle classes, that opinions and theories would have to be judged by recourse to evidence and reason so that people would be able to distinguish between reason and unreason, sanity and insanity, innocence and criminality.

The first thinker to moot this reorientation from reality to society was Hegel. (Though it has to be stressed that it did not appear to Hegel to be a reorientation because of his idealistic belief that reality and reason coincided!) One of his most astute contributions to philosophy was his observation about the functioning of self-consciousness. When a self-consciousness is totally alone and able to make no longer what in our terminology would be a wild hypothesis about its identity or content, it is totally subjective and unsure of itself. When two self-consciousnesses meet face to face, Hegel continued, they confront one another in a struggle of life and death. Each, sticking to its own hypothesis about itself, wills the death of the other. Put less dramatically, Hegel meant to say that when two hypotheses about a state of mind conflict, there can be no way of solving that conflict except by the elimination of one of the hypotheses. The conflict becomes a battle of wills; and the stronger will subdues the weaker will. This is also the case when one takes a kinder view of the relationship between two persons. Suppose that a relationship is not a battle of wills, but a loving relationship. Even in this case, the agreement produced cannot provide assurance that the state of mind in question has been identified 'correctly'. It will simply be identified by loving consensus. But there can be no assurance as to what it is in itself.

When a third person joins, there arises the possibility of a non-agonal resolution of the confrontation. With three people, there comes the first possibility of an agreement between two and the outnumbering of the third. A majority decision can now take the place of the life and death struggle. Moreover, one can even identify the rules by which a majority decision can supersede the life and death struggle so that the struggle can be replaced by rule following.[6] Hegel's argument depended on his realisation that the individual identity of a self-consciousness could not be established by a close observation of that self-consciousness. Neither personal introspection nor the observations made by an outsider, he said, can yield an assurance as to what that identity precisely is. In modern terminology we would say that no ostensive definition of self-consciousness is possible, but that we must have recourse to rules of sociability. Instead of looking at reality, at what the self-consciousness really *is*, we identify it by the rules which obtain in the community we are members of. When three persons are involved it is possible for a rule, supported by at least two against one, to establish itself. By following that rule, a label or symbol can be attached to a state of mind; and whatever that state of mind is thus identified as, will stabilise itself as the 'correct' label or symbol.

In the middle of our own century, Wittgenstein proposed a nearly identical solution to the problem of reference. We cannot establish the object or event a sentence refers to, he argued, by pointing at the object or the event and then nodding and saying: 'There! This is what I mean!' The meaning of sentences, he insisted instead, is established by the rules which govern their use in a given community, not by their relation to reality. This argument amounted to a naturalistic treatment of the meaning of our sentences. Instead of setting up norms as to when and how reference could be proved to have occurred, Wittgenstein stressed the natural, social practices of man. This naturalistic anthropocentricity in Wittgenstein's thought was not immediately grasped. It was shyly referred to by both Strawson and Malcolm in their reviews of the *Philosophical Investigations*; but Feyerabend, in his review, did not comment on it.[7] However, Feyerabend did mention it in a subsequent article of 1954 in *Merkur*. The first author to bring it out in full was P. Winch in his *The Idea of a Social Science* of 1958. This social naturalism proved a powerful influence on Thomas Kuhn and Richard Rorty in America and on Karl-Otto Apel in Germany. It was critically discussed by Ernest Gellner.[8] It is mentioned in D. Pears' *Wittgenstein* of 1971 and in the same author's book on Wittgenstein, *The False Prison*.[9] It figures in S. Kripke's *Wittgenstein on Rules and Language-Games* of 1982,[10] and was finally summed up by G.H. v. Wright, who wrote that Wittgensteinian language-games are embodied as *Lebensformen*.[11] David Bloor made this naturalism and the consequent socialisation of knowledge in Wittgenstein the theme of his *Wittgenstein: The Social Theory of Knowledge*.[12]

Wittgenstein, in short, had implicitly and unobtrusively anticipated Quines's suggestion that epistemology ought to be naturalised. The only difference was that where Quine proposed to look towards the science of psychology with a little physics thrown in to explain neuronal activity, Wittgenstein was looking towards social institutions and, in so far as language-games are embodied as *Lebensformen*, towards our knowledge of the *Lebensformen*, that is, towards the science of sociology. As one of his followers puts it: 'What determines whether or not a concept properly applies to its next putative instance? The suggested answer is that a collective contingent judgement fixes proper usage.'[13]

Although the target of Wittgenstein's argument was the problem of reference rather than the problem of self-consciousness, the structure of Wittgenstein's argument is almost identical with the structure of Hegel's argument. In fact, since Wittgenstein rejected not only the possibility of establishing reference by ostensive definition, but also the possibility of establishing reference by stating what was in the mind of the author of a sentence, the difference between Hegel's and Wittgenstein's concern is not very great. Between them, Hegel and Wittgenstein redirected philosophical attention from a preoccupation with the information that comes from within and the information that comes from the front, to the preoccupation with the information which comes from behind, that is, from the rules of social solidarity. With such redirection, society moved into the centre and sociology tended to replace epistemology. The knowledge we harbour came to be seen more and more as a function of our social relations rather than as a function of a special epistemic relationship with the world the knowledge purports to be knowledge of. Community and solidarity came to assume an epistemic function and philosophers began to think of knowledge as the result of our membership of an epistemic community or solidarity. The point is well made, by Peter Berger and Thomas Luckmann,[14] but these authors are vague as to the reasons for the need to look towards epistemic communities. They vaguely speak of 'anthropological' and 'biological' reasons because 'interiorisation' is not possible. My argument is, on the contrary, that there are very specific reasons why philosophers have had recourse to the concept of epistemic community and that these reasons are wound up with the inability to *exteriorise* one's experience with any degree of precision or confidence.

There has always been a *prima facie* case for such a move. Whatever standard or foundation of knowledge we happen to have chosen – intuition, sense-data, authority, tradition, revelation, hearsay – we can never avoid the ultimate question as to *whose* standard and *whose* foundation. So, even in quite traditional attempts to explain knowledge and assess its validity, there looms in the end a question as to the social status of the people whose sense-data or revelation we accept. Clearly, when sense-data are involved, we do not regard those of the blind and the insane as relevant.[15] And when revelation is involved, we have sometimes taken care, or ought to have taken more care, that psychopaths like Hitler are disregarded. The decision as to whose sense-data and as to whose revelation is to be excluded is a social decision and depends on socio-political power, not on cognitive factors. For this reason even in traditional epistemology, a certain amount of sociology has never been out of place. But whereas in conventional epistemology such sociological considerations were peripheral, with Hegel and Wittgenstein they have moved into the centre of attention, and Michel Foucault is now insisting that all allegedly cognitive decisions must be admitted to reduce to questions of socio-political power because the determination as to who is insane and who is sane, who is criminal and who is not, cannot depend on cognitive factors, but is made according to political considerations, mainly of power, applied by bureaucrats and the police.

III

The search for non-cognitive determinants of knowledge began during the eighteenth century, almost at the same time as the growth of knowledge really got under way. We are used to thinking of the Age of Enlightenment as the century of scientific growth and as the century in which knowledge became scientific because it managed to extricate itself from religion. But if one takes a closer look at the Enlightenment, one will readily see that it was an age which stimulated scepticism not only in regard to traditional religious beliefs but also in regard to any knowledge, including the new scientific knowledge. 'The philosophers' glorification of criticism and their qualified repudiation of metaphysics make it obvious that the Enlightenment', Peter Gay wrote,[16] 'was not an age of reason but a revolt against rationalism.' The Enlightenment sowed doubt not only in regard to religion but in regard to all knowledge. The reason why this point is so frequently overlooked is that we have taken our assessment of the Enlightenment from Voltaire, its greatest purveyor. Voltaire was something of a historical 'catastrophist' long before the geological catastrophists had established themselves. Voltaire believed that the Enlightenment had suddenly broken upon mankind and illuminated it by its rays like a sudden 'catastrophe'. He believed that its greatest protagonists were Locke and Newton and that its emergence had taken place in England. It was there that the medieval world of superstition had suddenly collapsed. Voltaire was the first to introduce the reading public of Europe to the marvels of England's urbanity and civility, to the generous infrastructure of London, to coffee-houses and newspapers, theatres, assembly rooms, philosophical and royal societies and lending libraries. In this atmosphere, he told his readers, the progress of rational knowledge must be assured and the rest of mankind would soon benefit from it. In his view, the battle against religious superstition was led by the progress of science and reason.

The picture painted by Voltaire was basically false. There is no space to go into details. A sketch must suffice. The methodology advocated by Locke could not be applied, least of all by Newton. Newton himself was frequently confused as to how he had gone about getting his marvellous knowledge and knew that he had developed one strategy for gravity and another one for optics. A dispassionate and non-Voltairean look at the eighteenth century reveals widespread scepticism. Indeed, the whole rise of Romanticism in that century stemmed directly from that scepticism and was therefore very much part of the Enlightenment's scepticism about knowledge rather than the antagonist of the Enlightenment's knowledge. Goethe, in lapidary poetry, declared in the opening verses of his *Faust* that we can know nothing and that that knowledge was burning his heart out. A little later, his hero declares that 'in the beginning there was the deed', not knowledge and not reason, and that it was the deed from which everything else stemmed. Rousseau and Wordsworth, in obedience to such scepticism, had upheld the voice of the heart and extolled the simplistic naivety of children against the sophisticated criticism of reason and knowledge. Hume's scepticism about Newton was quoted above in Chapter 2,[17] and even Voltaire in his more reflective mood, very much like Goethe himself, had grave doubts about knowledge and recommended that we cultivate our gardens instead. If life is a disaster area, these men seemed to be saying, we can transform a little acre of it into a garden through hard work and by a deed – but not through knowledge.

Whichever way we turn, we find that the most enlightened part of the Enlightenment was not the growing confidence in the power of reason and science, but an increase in scepticism. This scepticism came in many different guises. In Hume it was specific and articulate as far as knowledge was concerned; and tinged with pessimism, as far as morals and social standards were concerned: not reason, he said, echoing Shakespeare, but passions and appetites govern human behaviour. In Goethe, this scepticism in regard to knowledge developed from despair[18] to a robust pragmatism which placed deeds before knowledge. With Romantics from Rousseau to Wordsworth and Keats it tended to a positive dislike and even contempt for knowledge and extolled the purity of the heart's desire at the expense of knowledge. All in all, the upshot of the Enlightenment was a down-grading of the cognitive factors in life and, depending on one's temperament, the emergence of a stoical scepticism, a despairing resignation or a glorification of the heart and its passions. Remarkably, the Enlightenment of the eighteenth century managed not only to introduce knowledge and science, but also to stand on its head the old tradition in which the passions had been supposed to have been chained by reason and in which action had been supposed to have been dictated by knowledge – a tradition that went right back to Plato.

But philosophers would not concede defeat. From Frege onwards, there is a long succession of efforts to define the meaning of meaning, the terms of reference and the conditions of truth – all efforts designed to take the sting out of the Enlightenment's scepticism and to show where and how our knowledge refers to what is known. Such attempts have become one of the chief preserves of twentieth-century philosophy. The type of all these attempts is Tarski's famous definition of truth: a sentence, he said, is true if what it says is the case. 'It snows' is true if it snows. This is logically quite impeccable, but, of course, totally meaningless as an answer to Hume or to Goethe, because it studiously leaves the great question as to how precisely we determine whether it snows, unanswered. It not only leaves it unanswered, but it even creates the impression that the answer does not matter. By sharpening the logical precision of the theory of truth, one creates the delusion that an actual gain has been made, when in reality all that has been achieved is that the heart of the matter has been relegated to a nether region in which all attention is concentrated upon logical and semantic finesse.

Given the delusion, there was then every incentive to pursue the search for the social basis of knowledge and to elaborate on it, rather than to play it down. The more philosophers strained to define reference and truth, the greater the temptation for sociologists to come up with a sociological explanation of knowledge. The idea that knowledge does not come from a relationship between man and the world but from a relationship between man and man developed into a pursuit which was both facilitated and encouraged by the new understanding of society which had been making itself broad ever since the eighteenth century, that is, at the very time when the deficiencies of knowledge and scepticism in regard to cognition had forced themselves to the fore.

The new conception of society which countenanced this emerging alliance between philosophy and sociology consisted of three different parts. First there was the growing belief that nature is something very closed to us because it is inert, incapable of speaking, made by God, and if not by God, certainly not made by man. Things that are not made, Vico had argued, cannot be known by man. Humans can only 'know' what they themselves have made because they themselves then know what they made it for. While society may not be fully man-made and may not even appear to be fully man-made, it does appear a great deal *more* man-made than fire, water, the earth and the heavens. Hence the turn towards the study of society, unlike the study of inert nature, held out a promise. The application of the Vichian principle was certainly encouraged by the thought of those philosophers who believed that society owed its origin and its dispositions to a social contract, i.e. to articulate human intention.

Second, the growing preoccupation with the study of society had shown that the old classificatory scheme of monarchy, oligarchy, aristocracy and democracy was insufficient because it had looked at social institutions entirely from the point of view of the distribution and exercise of power. There was a great deal more to society than power. People were realising now that social institutions themselves owed a lot to climate and geography and that the distribution of power in every society was merely an expression of these other factors, and certainly not the determinant of social customs. The domain of the social, it came to be recognised, extended beyond power and politics, into commerce, the arts, possibly even into religion and most clearly into private morality. The novels of Stendhal are replete with comments to the effect that a certain emotion, a certain behaviour, would have been different had it taken place in Paris and that Parma dictated an erotic passion which would not have flourished in a village in the environs of Besançon. To someone thinking along these lines, it was becoming clear that the category of the social had ramifications far wider than anything dreamt of in Plato's *Republic* or Aristotle's *Politics.* Although in the eighteenth century nobody had yet taken this step, one could see how the new thought about society invited the possibility that knowledge might be included among the many factors which were determined by social arrangements. Nor did the development of Burke's anti-rationalist, anti-contractual thinking about society detract from the understanding of society as encompassing commerce, private morals, the arts and, possibly, even knowledge itself. It encompassed the possibility that commerce and the arts as well as private virtues were determined by social tradition and could be taken to mean that knowledge too was the more valuable the more tradition-bound it remained. Hence

Burke's phrase that 'prejudice' gives real patina to knowledge was not intended to put knowledge down, but merely to establish its link with social order.

Last but not least, there was the experience of the French Revolution. When the Revolution began, people who were not carried away by enthusiasm had shaken their heads because they predicted the end of all social order. When the Revolution had been and gone and social order had not vanished from those parts of the earth that had been affected by it, thoughtful people started to learn a lesson. They began to realise that societies are not identical, let alone synonymous, with the particular order and power structure they happen to exhibit at any one time. The divine right of kings and the hierarchy of estates came to be seen as only one of many possible surface phenomena. Underneath, people began to grasp, there is a deeper structure which will force up a succession of different surface orders, so that one can be quite confident that societies survive even if their temporary surface order is changed or abolished. The question which then came to be asked was what the deeper structure consists of and why it is not affected by the demise of the surface structure. At this juncture the inquiry which we have come to know as sociology was started. The studies of history and of law, the more traditional methods of knowledge of societies, had only concerned themselves with the coming and going of the surface appearances. Now a new question moved into the centre. This was the question as to what social institutions or habits there must be for societies to endure in spite of the passing of any particular surface appearance of social order. As a methodical science of society distinct from the study of law and history, these inquiries were slow to start. But the word 'sociology' and the idea that there was a sociology of societies gained ground from the time of the Restoration onward. The heart of this idea was that there must be rules and norms of social behaviour which were distinct from the laws which prevailed, say, in a monarchy, or in a medieval city-state and which would continue to operate and enable social structures to change their laws and institutions. The idea that a social order is always and above all a rule-bound order irrespective of the specific content of any of these rules was to have a great future in philosophical thought, because it was this idea more than anything else which was capable of filling the gap created by the conclusion that all reference must remain inscrutable and all identification of states of mind, hypothetical.

IV

The thrust towards the derivation of knowledge from sociological rather than cognitive factors was quite slow to gain momentum; and when it did, it did so in three distinct movements of thought which were not complementary to each other.

The first move was made in an emancipatory intention and concerned itself exclusively with false knowledge. It was argued in the Introduction that false knowledge is adaptive because it has the special power of serving as a foundation of a social order. It can provide a bonding principle in the sense that all people who espouse a specific piece of false knowledge can consider themselves as members of a society. That society is bounded as well as bonded by subscription to that specific piece of false knowledge. What is more, only false knowledge can act as a social foundation because true knowledge could be espoused by anybody at all and could therefore never be used to set some people apart from others as a designated society, to define an exclusion principle and to provide that most important principle for the preservation of any one social order, the principle which defines where altruism must stop.[19] In this way one can distinguish along a broad line between true knowledge and the sum total of all systems of false knowledge. The latter can be used as foundation charters for societies; but the former cannot. This way of distinguishing between false and true knowledge is very broad and not of much practical use; nor is it possible on the basis of this distinction to offer cogent criticism of false knowledge. Since false knowledge has a clear adaptive value because it provides the possibility of being used as a foundation charter for a social order, it promotes solidarity and cooperation. As a result it has to be treated with respect, even though such respect comes from a motive very different from the respect meted out to true and genuine knowledge. If emancipation from false knowledge is required, it will be necessary to indicate the specific reasons why any particular set of false knowledge is ever harboured and promoted, so that, with the elimination of those reasons, the particular set of false knowledge can be given up.

Karl Marx

The first thinker who systematically linked false knowledge to specified social institutions or movements was Karl Marx. His primary intention was emancipatory because he wanted to liberate mankind from the spell of all sets of false knowledge. He thought he could best do so by establishing which sets of knowledge were linked to social institutions. The sets so linked must be sets of false knowledge because they owed their presence to factors other than cognitive ones. Marx's search for social determinants of knowledge was therefore confined to sets of false knowledge. In his own mind he held that the natural sciences, mainly physics and chemistry, but to a certain degree also biology, were known through and determined by cognitive factors and therefore not in need of a sociological explanation. The distinction between the natural and the human or moral sciences, however, cannot be hard and fast. Marx himself tried to draw the line of distinction by asking himself whether a certain body of knowledge affected human beings and their relations towards each other. In that sense one might say that astronomy was a natural science because it did not affect human beings. On the other hand, psychology is a system of knowledge which clearly has a direct bearing on human beings. In between these two extremes, there are the sciences of history, physics and chemistry. Their bearings on human life varies. With history it is obviously very great, with physics less so. But when chemistry and physics are being used as a basis of medicine, they start to impinge on human beings directly. And where biology, as in social Darwinism, was used to promote and justify social action and policies, it too had a direct bearing on human life.

Although there must, therefore, remain some doubt and argument as to what knowledge is to remain exempt from a search for a social explanation, the line of Marx's argument to promote emancipation from false knowledge is clear. He argued that that knowledge which can be shown to serve and promote the interests of the people who espouse it and propagate it is false knowledge, for it can be shown to be based on non-cognitive considerations. This argument by itself is acceptable and is not basically different from the argument, put forward above, that false knowledge can serve a social purpose and can, therefore, be adaptive. It was argued that false knowledge can become a basis for solidarity and cooperation of a whole group because it sets that group apart from other groups. Obviously this argument takes its cue from our knowledge of the practice of primitive tribes. But the same argument would apply to larger societies in which one segment, say the class of capitalists, is using a certain belief system to identify its members and to set them apart from other segments. It can also be applied to the situation in which such a belief system is used by one segment of a society in order to validate the superiority or hegemony that segment is enjoying over other segments of the same society.

The emancipatory strategy proposed by Marx becomes, however, highly questionable as soon as one considers the following situation. If there is an independent criterion for distinguishing true knowledge from sets of false knowledge, the Marxist strategy is in fact liberating. One can then simply address oneself to all sets of false knowledge and seek an explanation as to why, though false, they keep being used and promoted. As soon as one can isolate the social interest which makes people embrace a particular set of false knowledge, the liberation of the mind from such false knowledge can begin. But suppose there is no independent criterion for distinguishing between sets of false knowledge and true knowledge. In that case the success of the emancipatory strategy itself becomes the sole criterion. Wherever one can point to class interest as an ingredient in a set of knowledge, that set of knowledge is identified as false knowledge. Obviously this is not enough. Surgeons, for example, have an obvious interest in promoting a certain kind of knowledge, and they benefit financially from such promotion. It would, however, be a mistake to conclude from this relationship between their interest and their knowledge that their knowledge must be false. It is here that we can identify the untenable element in Marxist emancipatory strategy.

Marx himself, on his own showing, believed that he could provide a practical criterion of true knowledge – that is, knowledge dependent entirely on cognitive factors – by his distinction between the natural sciences and knowledge that affected human beings. But we have seen above that that distinction is tenuous at best and that even the science of astronomy, which in his days must have been totally beyond the reach of human beings and therefore, by his criterion, totally exempt from the strictures that must be applied to other sciences, has, as a result of advances in physics and chemistry, moved very much into the orbit of those sciences which have a bearing on human affairs. The mere fact that this situation has changed so much since Marx's days indicates that Marx's own distinction was unstable and probably quite invalid. Sooner or later almost anything could move into the orbit of knowledge which affects human beings. If, in Marx's own day, astronomy might be indisputably beyond relevance to human lives, there were many other disciplines the objective standing of which was even in Marx's own day already in dispute. Marx, for example, believed that the theories of Malthus were an ideology (i.e. false knowledge) espoused by the aristocracy because they served their class interest and would protect their privileged position. At the same time, he believed that the science of craniology was sufficiently natural not to be suspect!

On the other hand, Marx believed that the theories of Kant were ideological because in justifying individual judgement and conscience, they served to protect the class interest of the bourgeoisie, that is, of those people who by their education and income could afford to exercise their judgement and listen to their conscience. As to Darwin, Marx himself was actually in two minds. Darwin, he thought, had argued from the actual war of all against all in English society to a war of all against all in the realm of organic nature. But when Darwinists like Spencer argued that the war of all against all was a necessary part of all human societies, they were propounding ideology rather than knowledge, for their theory clearly served the preservation of the status quo. It is clear, then, that Marx did not really have at the back of his mind a criterion for distinguishing between true and false knowledge. He was therefore unable to deploy his strategy in a truly emancipatory sense against false knowledge. On the contrary, in the absence of a genuine criterion, he and most of his followers tended to use his strategy to identify all knowledge as false knowledge. For this reason, the entry of sociological considerations into the philosophy of knowledge was ominous, even though it was motivated by the best of emancipatory intentions.

Emile Durkheim

The next decisive and influential move in the transformation of the problem of knowledge into a sociological problem was made by Emile Durkheim in 1911 with the publication of *The Elementary Forms of the Religious Life.* Unlike

Marx, Durkheim was not primarily interested in emancipation. His purpose was to explain why, all over the world and as far as we can think back, people have always observed religious, as distinct from more ordinary, utilitarian or economic, practices. The explanation he gave was a great advance in knowledge for two reasons.

All sets of religious observances, with the possible exception of a few sets performed by exceedingly primitive tribes, have always come together with an explanation of why they were performed. The trouble with these explanations was that they were capable of explaining only those observances kept by the people who also advanced the explanation as to why they were kept. Thus the Hopi Indians would advance an explanation which explained Hopi observances. Christians would advance an explanation which would explain Christian observances. Muslims, again, had a perfectly good explanation of why they were keeping *their* observances. Durkheim's merit was that his explanation worked very well for all religious observances. In this sense it had to be preferred to Jewish or Muslim or Christian or Hopi explanations because it had greater explanatory power than any one of the other explanations.

Durkheim's explanation was, second, a real advance over a host of other explanations which had gained ground ever since the Enlightenment. These post-Enlightenment explanations were historicist. They did not so much explain as insist that the history of mankind was a gradual ascent from religious observances to scientific knowledge. None of the theories, from Voltaire and Turgot to Comte and Frazer, had bothered to add an explanation of why this should be so. It was simply taken as a historical fact that this gradual change of mental attitudes had taken place and that it was inevitable. Since mankind has only had one single history, one could not even argue that this historicist theory was the result of the observation that all mankinds on all conceivable planets had always developed along these or similar lines. This historicist theory could be given the strength of an explanation only by the addition that religious observances are the result of superstition and that superstition is the sort of false belief children engage in or are captive to; and that the earliest ages of mankind must be likened to the stage of infantile mentality. As any reader must recognise, these additions, which alone could make historicist explanations of religious observances into genuine explanations, are based on a host of unfounded opinions. It is by no means obvious that mankind should go through stages of development which resemble the mental development of a human being. Next, it is far from obvious that religious observances are simply the sorts of superstitions children are prone to. And even if all these additional opinions were to be defensible, we would still require further additional explanations as to why in the history of mankind there was this particular growth of mentality. Whichever way one looks at it, these historicist explanations have very little explanatory power because of the weakness of the auxiliary hypotheses required.

Durkheim's theory of religious observances was powerful and effective and effectively superseded the then current historicist explanations. It divided the entire universe into two classes of things which completely excluded one another. One class consisted of profane things; and the other, of sacred things. Sacred things have to be protected and isolated; and profane things must be kept at a distance from the sacred things. In this theory attention was shifted from the cognitive problem as to how one might distinguish between superstition and true knowledge to an inquiry into those social institutions which were for the most part concerned with the protection of the sacred and into those manipulations of the profane which would ensure that the sacred was not being polluted. This bipartite distinction, together with the notion that religious *beliefs*, far from being the foundation of these institutions, resulted from the manner in which social institutions reflect this bipartite division, corresponded with Durkheim's general philosophical outlook and underwrote it. Durkheim had no difficulty in accepting that the profane world impresses itself upon our senses and that all sorts of beliefs and common-sense propositions can be explained through some kind of sensualism. He was, in other words, for many practical purposes an uncritical common-sense positivist. But all those elements in our knowledge which cannot satisfactorily be explained in terms of simple sensations – concepts like time and space, cause and force – make their entry into our minds or into our conceptual apparatus via our experience of social institutions. His explanation of religious beliefs had proceeded from the observation that there was a simple dichotomy of social institutions. Similarly, he explained the fact that we think in terms of time and space and in terms of causes as the result of our experience of social institutions. In Durkheim's world, social institutions were primary and simple; and could be used to explain all those phenomena which were not primary and simple. All those things which are not due to our senses are to be understood as socially determined, imagined representations. For good measure, he could thus also explain why such concepts as cause, space and time, though not derived from our sense-experience, are used universally. They are so universally used because they derive from social institutions which, in one form or another, are completely ubiquitous.

He applied this reasoning to quite specific concepts. He argued that the concept 'kind', without which we could neither classify nor abstract, reflects our awareness that we are all members of a family group and that all members of that group have certain likenesses. Similarly, the category 'totality', he argued, must be derived from one's experience that one is a member of a group. The notion of space, which, he said, could not possibly be abstracted from limited amounts of sense-experiences, can be explained as a reflection of the experience that in a society, institutions are separate from each other. Social organisation is perceived as spatially extended and the category of space is a continuation of one's perception of social order. Similarly the concept of time is based upon an experience of the extension of social institutions. Durkheim argued that the notion of time is very different from individual experiences of memory and could not be derived from them, because 'time' is a universal category shared by all people, a universal order of succession which constrains all minds. If these categories are derived from profane social institutions, there are others which are an extrapolation from those social institutions which set activities and objects apart and thus define the order of the sacred. The power of the group is, for example, expressed as a sacred, impersonal force and gradually filtered into the abstract idea 'force' which pertains even to non-sacred events, e.g., the force of gravity. 'Causality' is a notion which is consequent on 'force' but requires the notion of 'necessity', a notion which in turn requires the notion of regularity, which is directly derived from the social institution of law, of the punishment of law-breakers and of the sanctions which are imposed upon the violation of law. Durkheim also attributed the very notion of 'truth' to our social experience of the impersonality and stability of the group we are members of. It is through social consciousness, he maintained, that the individual receives his or her first intuition of the concept of truth. In short, if man were to rely entirely on his individual perceptions and personal sensations, the whole edifice of knowledge, scientific as well as religious, would remain inexplicable. Man, he surmised, would then be indistinguishable from beasts. This last observation is ironically perverse. Durkheim was obviously completely unaware of how much knowledge is embodied in every single biological organism, from the single-cell bacteria upwards. He was therefore unable to entertain the idea that there are sources of information other than those either directly attributable to our sense-experience or mediated by our membership of social groups.

Durkheim's approach has been fruitful and has yielded a rich harvest in the course of the twentieth century. All the same, in fairness to Durkheim, one has to stress that he applied his sociological explanation of the categories which play such a crucial role in our knowledge only to those categories or parts of our knowledge which could not be readily explained by sense-experience. As time went by, confidence in positivism and in its explanation of all residual parts of knowledge in terms of sense-observations wore thinner and thinner. For this reason, many Durkheim followers widened his recourse to sociology and tended to universalise it. Inspired either consciously or unconsciously by Durkheim, many people, instead of seeking for alternatives to an increasingly discredited positivism, kept looking more and more towards sociology for an explanation of why we have certain beliefs. Thus, for example, Mary Douglas has become very famous for explaining that pollution fears are the result of a given social order and that they either disappear or would be directed to different objects if that social order were to be changed. It is true that there are some pollution fears, like the fear of menstrual blood or the objection to pig meat, to beef meat or to other perfectly nourishing foods, which have no rational basis. In such cases a Durkheimian explanation must always be welcome. But there are pollution fears which are very real and rational because they are cognitive, like the fear of lead poisoning or the fear of nuclear radiation. Any attempt to explain such fears by a recourse to a Durkheimian sociological argument appears, therefore, completely redundant. The ease with which people have advanced such explanations in spite of their obvious redundancy and the popularity which such sociological explanations enjoy is a cautionary tale and goes a long way towards a demonstration that one should have recourse to Durkheim only as a last resort, when all other explanations have failed.

The Marxist strategy of changing questions of knowledge into questions of social organisation tended to produce the result that all knowledge seemed to be false as soon as it was seen to have been a method of class exploitation, regardless of whether it had also been a method for understanding and explaining nature. The Durkheim strategy tends in the opposite direction. The more knowledge is explained as a function of social institutions rather than as theories about nature and society,the more one ends up with the view that all knowledge is true knowledge. Since all human beings are living in societies, all knowledge which can be explained as a function of a social order must turn out to be true knowledge. We come here to the slippery slope of the sociology of knowledge. One of the prime requisites for any theory of knowledge must be the ability to explain the differences between true knowledge and false knowledge. If we go along with a sociology of knowledge which explains all knowledge as false knowledge or if we go along with a sociology of knowledge which explains all knowledge as true knowledge, we are abandoning the project we were embarked on – which was a project to tell us how to distinguish between false knowledge and true knowledge when reference neither to reality nor to authorial intention proves feasible.

Voltaire, Ranke, Spengler and Malinowski

The third move in the direction of the sociologisation of knowledge was made in the middle of the twentieth century by the work of the later Wittgenstein. I say 'later' advisedly, because this work was largely undertaken by Wittgenstein in reaction to the shortcomings of his own earlier *Tractatus*. All the same, the idea that knowledge is primarily a sociological rather than an epistemological problem had a venerable ancestry – an ancestry which one can trace back at least to Voltaire in the eighteenth century and which had a lot to do with the observation that it is easier to explain a body of knowledge as a system which is self-contained than as a set of theories and propositions each one of which refers to something outside that system. Voltaire's innovation in this direction was simple and quite unpretentious. It had occurred to him during his historical work and had remained confined to his understanding of history. Historians before Voltaire had been wont to explain any epoch or age largely as the result of earlier events. In this method, whatever happened in the epoch they were writing about had to be referred causally to events which has preceded that epoch. By contrast, Voltaire formed the fruitful idea that one could meaningfully dispense with that kind of diachronic reference and simply explain every epoch like the Age of Louis XIV as a self-contained system. Incidentally, he was encouraged in this method by his idea that the history of the past was a history of superstitions and moral depravity and ought therefore to be allowed to fall into oblivion. With this view he had a prime incentive for disregarding the reference earlier events might have had to the age of Louis XIV. Instead he made the epoch he was then writing about appear like a closed circle in which military history could be referred to synchronic art and art to synchronic political institutions and so forth. Voltaire was the originator of this special brand of cultural history, which was to find its most famous practitioner in Jacob Burckhardt in the middle of the nineteenth century. The idea of treating a set of events as a closed circle is fruitful and had, given the difficulties one encounters with the concept of reference, so much to be said for it that it was taken up time and again even though there is no reason to believe that the other people who took it up were conscious imitators of Voltaire or even accidentally influenced by him.

During the nineteenth century this same idea appeared in the work of the German historian Leopold von Ranke, who conceived the notion that the entire history of mankind is the sum total of closed circles, known as nations. Each nation, he taught, and here he was consciously following the ideas of Herder, was informed by its national spirit or its national genius, which could be used as an explanatory principle of all the institutions it gave rise to and of all the developments these institutions had undergone. Each of these nations, he surmised, must be 'equidistant from God'. This pregnant, theological metaphor was designed to express the view that each nation was not only a closed system, but that the ethical and aesthetic values it enshrined, sported or gave birth to, were at one and the same time relative to that nation and an absolute finality, incapable of further explanation. They could not be compared to one another or arranged in a descending or ascending order. Each nation was immediately created by God, just as Linnaeus had taught that each species is an idea in the mind of God. Whatever explanatory power these closed circles had, they precluded among other things all thoughts of evolution, for in evolution either the earlier or the later is nearer to God or nearer to whatever point of departure one supposes evolution to have started from. Though Ranke was a historian and this view of closed circles came to be called 'historism', it is justifiable to extend this term to all closed-circle methodologies in which the closed circles do not stand in an intelligible or causal, let alone evolutionary, relationship to each other.[20]

The technique of historism became the corner stone of two entirely different systems of social science which were propounded and widely discussed during the years immediately after the First World War. The idea is so powerfully persuasive, not to say self-explanatory, that its re-emergence in the twentieth century owed nothing or nothing much to either Voltaire or Ranke. It was used as the general explanatory principle of the seven cultures of human history by Oswald Spengler in his famous book *The Decline of the West* and by Bronislaw Malinowski in his renowned work on the Trobriand Islanders. Spengler made spectacular use of the idea that whatever happens inside a single culture has to be understood as a function of the central paradigm of that culture and does not refer to anything that happened earlier and cannot be seen as the cause of anything that happened later. In Spengler's view each of the seven cultures produced by mankind was a self-contained whole. Even its rise and decline, he argued, could only be explained as the result of an inner biological law of growth and decay, never as series of causes and effects. The biological law of growth and decay was the same for every culture, whereas a series of causes and effects would have been likely to be different in every culture.

In Malinowski's view, the culture of the Trobriand Islands could best be explained as a self-contained functioning system in which Trobriand myths were functionally related to Trobriand economics and Trobriand economics to their family structure, and so forth. The notion that some of these Trobriand institutions could be explained by their history, i.e. by reference to a Trobriand or any other neighbouring past, was rejected by Malinowski out of hand. Malinowski's historism came to be known as 'functionalism'. In a functionalist explanation, reference to events and institutions outside the system are avoided and everything that happens inside the system is explained as a function of something else that happens inside the system. To put it in the words of Maurice Bloch:

> For Malinowski everything was to be found in the anthropologist's conversation with the people studied. Since the past is not to be *seen* in such conversation, it has no explanatory value; and, on the other hand, when it is talked about, it has to be explained in terms of the present.[21]

Malinowski's method seems to have anticipated the later Wittgenstein's insistence that the meaning of sentences is functionally determined by the rules of a community and not by reference to something outside like an ostensive definition or a reference to its author's personal intention.[22] Spengler's historism was either taken for granted by the intellectual milieu to whom he had addressed his work; or was considered *sui generis*. Malinowski's historism, on the other hand, was developed in explicit reaction against the then widely current anthropologists' historicism.[23]

Ludwig Wittgenstein

The strategy of closed circles was fully taken up by Wittgenstein at the time when he was beginning to find shortcomings in the positivism which had inspired his early work. Wittgenstein, in the end, took up a full-blown historism in regard to knowledge systems; but gave it a radical and completely original foundation. Already Durkheim, when he had been trying to find the locus of concepts in social institutions, had noticed that it was not only necessary to explain concepts like space or causality as such; but also necessary to explain that these and other concepts were held by all human beings and could therefore not be believed to have had their origin in individual observations. The need to provide this kind of explanation had led him to his view that these concepts must be collective rather than individual representations. The collective nature of these representations, he went on, could be more easily explained if one derived them from social institutions; for all human beings had a share in social institutions.

Whether influenced by Malinowski or Durkheim or not, Wittgenstein took up the idea that the meaning of concepts cannot be a matter of individual decision and then proceeded to present it in a way which went to the heart of the matter. Though it has been alleged that Wittgenstein did little more than rediscover Durkheim,[24] there is a great deal of difference between thinking with Wittgenstein that we are following rules and thinking with Durkheim that we are constrained by rules of which we are not the intentional authors. Either way, the individual is not the author of her or his behaviour – and this is the starting point for both Durkheim and Wittgenstein. Concepts, Wittgenstein said, are stated in language; and language cannot be a private matter. The very nature of language consists in the fact that it is a mode of expression that is rule-bound. People who speak a certain language do not make the rules. Nor do they hope to convey their meaning by making idiosyncratic vocal gestures. If they did, nobody would understand and the exercise would fail in its purpose. In drawing attention to the rules of all languages Wittgenstein ought to have found himself eventually on very solid biological and evolutionary ground and would have had to recant his own epigram that the Darwinian theory had no more to do with philosophy than any other hypothesis of natural science.[25] For if one looks at it from the point of view of evolution, one must encounter the rule-bound nature of language and understand that the idea that a language is a privately felt or conceived way of conveying inwardly held meanings is inconceivable. Where the idea has been conceived, as in the private-language schemes devised by James Joyce, discussed in Chapter 1, it can be seen to be an idiosyncratic failure. Animals capable of making noises or gestures have an adaptive advantage. But that advantage is only useful if these noises and gestures are made according to rules. Without rules they would remain private expressions of opinions or wishes or fears and could not be understood by their fellow-creatures no matter how emotionally satisfying such expressions might be to the individual concerned. In other words, the adaptive advantage of making noises and gestures depends on the ability to make them according to rules so that there can be repetitions and so that these repetitions can, by natural selection, be grasped by other animals of the same species. Without Darwin we could form no appreciation of the importance of rules in language; nor could we understand that the following of rules is not an accidental but an essential characteristic of speaking a language.

In seizing upon the heart of the matter, Wittgenstein transformed historism from a methodological convenience into a philosophically powerful argument. The meanings of words, he said, do not depend on the private intentions of their users; nor, he continued, do they depend on an ability to point ostensively to the events or objects to which they are alleged to refer. There are no private languages and no ostensive definitions. The meaning of words and sentences depends, on the contrary, on the use they have in a particular speech community, that is, on the language rules in force in that community.[26]

In the first instance this doctrine does not say anything specific about the truth of sentences. It concerns itself exclusively with the meaning of sentences. But with this doctrine the truth of sentences is affected at the drop of a hat. For if the meaning of sentences depends on the rules in force in any given community and if the meaning of these sentences could be changed if they were pronounced by members of a different community according to different rules, obviously, their truth would be affected. Suppose somebody says 'it is raining'. In the community in which 'raining' refers to raining, the sentence is true or false according to whether or not it is raining. But suppose the same sentence is uttered by somebody who belongs to a community in which this expression is taken to mean 'I do not feel well': the truth or falsity of the sentence would immediately cease to have anything to do with rain. In short, by going to the heart of the matter and in saying that the meaning of expressions depends on the speech rules in a given community, Wittgenstein implied that sentences do not have invariant meanings which might depend on their reference to events. He implied instead that all meanings vary according to and with the speech community in which they are used. This doctrine has become known as the doctrine of the denial of meaning invariance.[27] If all meanings are variant, all meanings must be relative to a given speech community; and if all meanings are so relative, any truths they enshrine must also be relative to that community.

Wittgenstein here rushed forward where Marx and Durkheim had barely feared to creep. In order to emancipate the mind, Marx had drawn attention to the fact that there are truths which obviously serve the interests of a ruling class and which must therefore be suspect, if not downright false. Interestingly, Nietzsche soon after had stood Marx on his head and argued that there are beliefs which must be false because they so obviously serve the interest of the oppressed classes – e.g. the belief that the meek shall inherit the earth. Durkheim, in order to explain how knowledge makes use of concepts which cannot be said to be based on the observations of individuals, had tried to show how such concepts are collective representations of social experiences rather than of the facts of life or of nature. Wittgenstein turned out to be more intrepid than either of these two thinkers. He went to the heart of the matter and demonstrated that the very use of language is bound by rules. Since the sets of rules differ from community to community, there can be no meaning invariance of sentences or concepts. And then he concluded inadvertently but, to my knowledge, never explicitly, that if there is no meaning invariance, there can also be no truth other than a truth accepted in a community. Wittgenstein's language-games, David Pears observes,[28] act as self-correcting systems because we correct one another as we exchange speech, and so 'conformity is enforced by the need to communicate'; and he adds that 'the correctness of a practice can be judged only by its own internal standards ... Where else could we go for answers?'[29]

We have seen that with Marx it turned out that almost all knowledge was false; and with Durkheim, that almost all knowledge was true. Now, with Wittgenstein, we find that it no longer matters whether knowledge is true or false. As long as it can be shown that it is couched in sentences that obey the rules of a speech community, the knowledge conveyed by those sentences must be meaningful knowledge in that community.[30] 'A blunder', Wittgenstein said, 'is always a blunder in a particular system: just as a blunder in a game is always a blunder in a particular game and not in another.'[31]

With Wittgenstein, the sociologisation of knowledge reached its peak. Whether the acclaim with which Wittgenstein has been greeted would have been less wide if the ground had not been prepared by Marx and Durkheim as well as by other historists is very difficult to say. Soon after Wittgenstein the sociologisation of knowledge found its triumphant apostles in Thomas Kuhn and Jacques Derrida.

Thomas Kuhn and Jacques Derrida

It is not difficult to see what a historian of science like Kuhn and a literary critic like Derrida have in common. The strategy they share is the strategy of treating the subjects of science and of literature as if science and literature formed a closed circle and did not refer to anything outside the circle or, if it did, as if such reference had no bearing on its meaning and truth. Whatever the differences between science and literature – between a scientific theory about the world and a literary description of the world – might be, Kuhn and Derrida treat the scientific and literary descriptions of the world as if each could be assessed and judged entirely without reference to the outside world. Both Kuhn and Derrida maintain that what purports to be a description of the world is in reality a closed system presided over for Kuhn by an informing paradigm and, for Derrida, by a hermetically sealed text. If one follows their argument, a scientific description is not to be judged by its truth, that is, by its correspondence with the world, but by its coherence with a prevailing paradigm. Similarly, a work of literature is not to be assessed by the truthfulness of its description of the world, but is to be taken purely as a text which does not refer to anything which is not part of the text.

Obviously, there is a lot to be said for this view of scientific theories and of literary texts. The verification of theories, it has long since been agreed, is impossible. In so far as reference and meaning depend on verification, scientific theories can have very little meaning. The case which has been made for falsification as a criterion of the meaning of scientific theories is better; but not good enough, because any theory can be immunised against falsification both by *ad hoc* hypotheses and by an insistence that any particular falsifying instance was due to faulty observation or spurious experiment. Whichever way one turns, Kuhn's crucial insistence that an act of judgement which leads a scientist to reject a theory is always based on something more than a confrontation of the theory with the world, leads straight to the conclusion that above all a theory depends on the paradigm which informs it rather than on its direct correspondence with the world it purports to describe. Derrida's case is even more plausible, for a work of literature only rarely purports to describe the world. It is in a very real sense a work of fiction. It is made up by the author's imagination and, therefore, its actual correspondence with anything that has happened in the world is not and cannot be a criterion of its worth, its truth or its value. Nobody has ever seriously argued that the truth of a novel by Thomas Mann or by Marcel Proust depends on its correspondence with or reference to established facts. But it has traditionally been believed that its meaning depends on its ability to convey its author's intentions. In reading it, it is believed, do we or do we not manage to grasp what the author intended to tell us? But the problem of reference is intractable all round, and it is indeed impossible to establish that there is reference, whether it is to facts or to authorial intention.

It takes little perspicacity to detect the influence of Wittgenstein in both Kuhn and Derrida. Kuhn has freely acknowledged it. Derrida, as far as I know, has not. Nevertheless, the philosophical reasoning behind both Kuhn and Derrida is purest Wittgenstein. The meaning of a sentence, Wittgenstein said, depends on its use. The use, Kuhn explained, is controlled by an accepted paradigm. Meanings must therefore vary with paradigms, and every paradigm change entails changes of meaning. If the meaning of a sentence depends on its use in a given community, it cannot depend on the private intentions of its author, Wittgenstein had argued. If the meaning of a text, Derrida can be taken to have enlarged upon Wittgenstein, is not dependent on the intentions of its author, it must be ascertained by the interpretation of the text as an end in itself. For both Derrida and Kuhn, the meaning of sentences is to be derived from the text itself or from the paradigm itself. For neither of them can the meaning be dependent on or be determined by the reference of a sentence to the outside world. Truth and correspondence with reality, in other words, are not the primary determinants of meaning. This observation is not as unsound as it sounds. Suppose we wanted to detect the intention of the author of a text. All we could do would be to state that intention in so many words, thus creating a further text, the meaning of which would have to be established by discovering the intentions of the author of the second text; and so forth. The same reasoning holds for attempts to ascertain meaning in terms of any reference to events. One can only ascertain such meaning by formulating another sentence about these events, and one would then be left with the further problem as to what the sentences of the second text mean or refer to. The quandary we are in derives from the fact that we are inside the prison-house of language and cannot break out. Language, as Quine observed, refers to concepts; and concepts cannot readily lead us to a non-conceptual world, whether it is the world of inner, authorial intentions or the world of allegedly outside events. The advantages of the transformation of epistemology into sociology are quite striking.

V

With these advantages we arrive at a fully fledged sociology of knowledge. Such sociology explains the beliefs held and the knowledge harboured as function of a social order or a social habit. Even before the appearance of Kuhn's famous book in 1962, there were sociologists like Karl Mannheim in Germany and Robert Merton in America who had been thinking along these lines. Sociology, rather than epistemology, came to be considered the proper way to explain the phenomenon of knowledge and the kind of knowledge accepted as true.

Merton, for example, argued that the English Puritans of the seventeenth century believed that one must practise one's craft or calling with systematic and methodical industriousness. He concluded from this observation that these Puritan habits, rather than Bacon's reasoning in favour of experimentation, had promoted the practice of experiment as a foundation of knowledge. Bacon had urged that the men of the new learning should escape from Plato's cave and live in the contemplation of nature, as in the open air.[32] A Mertonian sociology of knowledge brushes aside Bacon's advice, and the possibility that it might have been taken seriously. Bacon, the argument goes, was thinking of nature, that is, of something it is difficult to refer to and of something other than society and must, therefore, have been wrong; and, being wrong, his advice could not possibly have been influential. If such arguments establish the sociology of the methods of science, other sociological arguments have been used to explain the content of science. Thus, for example, it has been argued by David Bloor that in the late seventeenth century the mechanical philosophy which insisted on the passivity of matter and held that particles do not move themselves and that such passivity was the result of a divine ordering of natural law was not based on observation and on reasoning, but designed to express Anglican latitudinarian social and religious ideology. The classification of things, it is alleged, reproduces the classification of people in society. If it seems irrational to explain both the method and the content of scientific knowledge as the result of social institutions, one ought to remind oneself that attempts to explain them in a seemingly more rational way as the result of observation and of reasoning about observation has proved well-nigh impossible. The sociology of science has a lot going for it! However, there remains one sobering observation. If traditional epistemologists have experienced difficulty with problems of reference to the world or to authorial intent, sociologists seem to be experiencing corresponding difficulties with their assessment of social determinants of belief. Where Merton holds that Newtonian mechanics were 'caused' by the social structure of seventeenth century England, A. Sohn-Rethel is of the opinion that the concept of Galilean inertia, which is so fundamental a category in Newtonian physics, was generated by the emergence of capitalism. Capitalism, he writes, necessitated the unending movement of money, and it was this notion of 'unending movement' which produced Galileo's concept of inertial motion.[33]

Apart from these disagreements about the substance of sociological explanations and a failure to show what kind of criticism can be used to eliminate at least some of these contradictory substantial claims, sociologists of knowledge also tend towards a methodological confusion. They readily speak of the social construction of knowledge. But they rarely make it clear whether they mean by this concept that sociology explains how knowledge is invented or whether they merely mean that sociology can explain how it is tested. Equally, they are ambiguous as to whether they think that sociology can explain why scientists and scholars accept it; or merely why people at large believe in it. It is equally worth mentioning that sociologists of knowledge are prone to reach for sociology too soon. When one is dealing with beliefs in witchcraft, sociological explanations are welcome; for, since women simply cannot ride on broomsticks, there is no rational explanation for the beliefs. But consider the case of geology. Martin Rudwick[34] distinguishes four styles of doing geology. Three of them – uniformitarianism, stratigraphy and scepticism – can be made credible by perfectly rational considerations and can stand on their own two feet. Only the last one, the style used by creationists and biblical fundamentalists, would seem to be in need of a sociological explanation. Rudwick, however, provides a sociological explanation in terms of Mary Douglas' grid-group variables for all four styles. The most he could claim for his analysis is that sociology can explain why so many people at any one time *embrace* now one and now another.

The distinction between the derivation of knowledge and the reason for embracing it as true is crucial. Foundationalist philosophers who believe that the method of discovery is the only conceivable warranty of truth fall an easy prey to the sociologists, because almost every discovery can be shown to have historico-social roots. It is, however, clear from what will appear in Chapter 4 that the method of discovery is not important and, indeed, is irrelevant. Just as volcanic eruptions or sun spots can determine biological mutations and genetic drift, so it is possible to find the socio-historical causes of discovery and even the psychological causes. But the method of discovery is neither here nor there. What matters is the test to which a discovery is subjected. The young Newton may well have longed for his mother and loved her at a distance and been prompted by this longing to think up the notion of action at a distance as Frank Manuel has suggested.[35] But it did not occur to him to advance this longing as proof of his thoughts on gravity.

It is true that even in the tests, historico-social conditions can be influential, as when Hitler believed the phoney evidence of the *Protocols of the Elders of Zion* to prove that Jews were engaged in an international communist plutocratic conspiracy or when the palaeontologists who explored the Burgess Shale were prompted by non-palaeontological beliefs to misinterpret their findings.[36] Here, too, sociology can be relevant, but not as relevant as it is to the method of discovery. And only foundationalists who are captive to the idea that truth depends on the method of discovery can welcome the sociology of knowledge. Those philosophers, however, who think that truth depends on the testing rather than on the method of discovery will relegate sociology to a more lowly place. There is a strange irony here. Foundationalists believe that the method of discovery is the only possible warrant of truth that is available. Next, they realise that discoveries are strongly influenced by socio-historical circumstances, and before long they are left with the conclusion that the theories which are believed to be true are the theories which have been determined by socio-historical circumstances and which therefore have, demonstrably, no reference to the world but assert something about the society the discoverer was a member of.

Let us take a closer look at the role of sociology in the testing of knowledge. Since reference is difficult to establish and since all observations have to be interpreted – 'sense-data are dumb', J. L. Austin used to say – a single person by herself must remain prey to doubt and uncertainty. Her judgement as to observed evidence – not to mention her judgement as to its possible relevance – will become stabilised only in the company of other people who will provide consensus and confirmation. Definition and determination of membership of such company and the relations of the single members to each other must always remain the province of sociology. It is therefore impossible to agree with Larry Laudan's conclusion – which he calls the arationality assumption – that 'the sociology of knowledge may step in to explain beliefs if and only if those beliefs cannot be explained in terms of their rational merits'.[37]

Next we come to the extension and the consequences of the sociological explanation of knowledge. If sociological explanations are accepted, it follows immediately that knowledge, even the most rational part of knowledge, science, cannot enjoy a special status. Knowledge and science explained sociologically are on a par with all other phenomena which can be explained sociologically, such as art, religion, common-sense beliefs, folk psychology, political ideologies and so forth. Next, it follows from sociological explanation of knowledge that all knowledge, including scientific knowledge, is relative to a paradigm, a language-game or a cultural system. There are, it follows, as many sciences as there are cultural systems. Even if one believes that beliefs and knowledge are dependent on perceptual equipment as well as on conceptual systems and standards of rationality, it follows from sociological explanations of knowledge that this equipment and these systems and standards vary from culture to culture. Since there is no independent way of deciding which of the equipment and systems and standards is best or better, we are left with total cognitive or epistemological relativism.[38]

The sociology of knowledge can therefore go a long way to satisfy the so-called 'Strong Programme'. (i) It can provide causal explanations of beliefs and states of knowledge. (ii) It can be impartial with respect to truth and falsity, rationality and irrationality, success or failure, of these beliefs. (iii) It can be symmetrical in its explanations, because the same types of causes would explain both true and false beliefs. (iv) It can be reflexive: its patterns for explanation will have to be applicable to sociology itself.[39] The last point is specially important, for it draws attention to the fact that sociologists of knowledge, while avoiding the problems of reference, are prone to wide disagreements about the social institutions and movements which determine belief and knowledge. There must, therefore, be a sociology of sociology: and such a programme invites an endless recess. This, however, is hardly surprising and not fatal. For, as Quine has often observed, if we make a naturalistic or sociological study of knowledge, then we can hardly expect that that study will yield firmer and more certain results than the science or knowledge it studied.

However, close examination of the Strong Programme will show that its four points are disparate. Only the first point is crucial. It shows that the entire enterprise is predicated on the assumption that what matters in knowledge is its foundations, its causes and its methods of discovery. If the methods are right, the assumption states, the results will be right. In Chapter 4 it will be shown that the foundations and methods are irrelevant. This demonstration will make the one crucial point of the Strong Programme superfluous. The remaining three points of the programme are trivial good sense, for one would expect a non-cognitive and non-referential explanation of the derivation of knowledge (e.g. 'Newton believed in gravity and held his theory of gravity to be true because, when he was a child, he longed for his mother, who was living in a distant village' rather than 'Newton believed in his theory because tests showed that its predictions were correct') to be impartial with respect to truth and falsity. One would also expect it to be symmetrical in its explanations and reflexive, in that sociology itself always stands in need of a sociological explanation. Indeed these three expectations are necessary and should not be seen as part of a new programme.

On the credit side, the substitution of sociology for epistemology solves the problem of the subjective mind and of how it relates to the world; and, more specifically, it solves the problem of how words (or minds) hook into things. In other words, it obviates the need for a solution of what we have seen to be the intractable problem of reference. Problems of meaning now become problems of hermeneutics and the meaning of a text can be entirely divorced from the authorial intention which guided the composition of the text, and so understood, the hermeneutic project obviates the need for a theory of reference. The advantages of this change are obvious enough. Given the difficulty of the alternatives, all of which wrestle with authors' intentions and questions of cognition which must always involve some recourse to psychology and thus to individual subjectivities, one has to be grateful for small mercies; even if knowledge understood and explained sociologically ceases to be knowledge *about* the world or an outside reality.

One would legitimately doubt that knowledge so explained is worth having. Systems of knowledge so explained are not cognitive phenomena, but social institutions. They are belief systems which support a ruling class or cement the bonds of a tribal group or whatever. But they do not contain information about the world. They can, at best, be seen to have some explanatory value of some phenomena even though they do not correspond to the outside world. For this reason pragamatist, instrumentalist, operationalist, and conventionalist philosophies of science often favour a sociological explanation of knowledge, because to them, a cognitive one is not available. But unless one is prepared to abandon all cognitive claims and regard knowledge simply as a social phenomenon with a social function, one is left with the ineluctable fact of relativism. For if what counts as true knowledge is determined by historico-social circumstances, all such knowledge must be relative to the historico-social circumstances which determine it.

VI

Some people are quite content to be relativists and consider relativism a mark of modesty and of fallibility – something like a humble tribute to the human condition – and, for this reason, relativists usually consider themselves morally superior. Relativism also enables Europeans to assuage their feelings of guilt about the havoc their imperialism has caused among non-Europeans. Others believe that relativism is self-refuting. They think that as soon as it has been shown that an argument commits one to relativism, especially to cognitive relativism, the argument has been refuted. If the modesty of the relativists seems arrogant as well as pretentious, the confidence of the second group needs qualification. For whatever its difficulties, relativism has a certain limited explanatory power. For example: in so far as people are living in societies, they are bound by rules and, therefore, able to appreciate the notion of abstraction. Members of a society are also able to grasp that, at least *within that one* society, there is an invariance of meanings and an absence of contradictions. For these reasons we must conclude that a sociology of knowledge and its consequent relativism can actually explain the fact that social beings are capable of a fairly high degree of rationality and that since such rationality can be found in other cultural systems, the view so dear to all relativists, that 'rationality' is a meaningless concept, is refuted. In this way, the sociology of knowledge not only has the advantage of obviating problems of reference, but also the ability to explain why there is a high level of universally valid rationality.

All the same, there remains a set of phenomena which it cannot explain. Social as well as cognitive structures are not all equidistant from the Big Bang, the primitive point of departure. Urban mass societies as well as highly general theories about the laws of nature are further removed from primitive states than societies of clans and the religious beliefs of neolithic tribes. Though there is no ground for believing that such lack of equidistance results from progress towards a goal, there is ample ground for thinking that this lack of equidistance has resulted from progress away from primitive conditions. While sociology can explain why we reject inconsistency and contradiction and why we are capable of abstracting regularities from a host of particular observations, it cannot explain why the movement has been progressively away from primitive states. Such an explanation can only be found in the following chapter, in which we move beyond sociology to an examination of cognitive conditions.

Even though there are some advantages in relativism, it can be shown to be an incorrect view. Bernard Williams put it in a nut-shell when he said that the assertion that all truths are relative requires a non-relativistic justification.[40] Relativism and the explanatory strategies from which it follows have one last and finally fatal flaw. For, since we are all living in the same universe, relativism is *ontologically* false. While it may be a good piece of temporary prudence, obviously it cannot be the last word.

The view that all truths and beliefs are relative to certain systems and have no application and validity outside that system is unable to explain how all these different, seemingly incompatible systems are related to each other. As they are all part of the same universe and coexist either synchronically or diachronically, there must be an explanation of their relationship. Any theory which fails to provide such an explanation must be a false theory. Now we have seen that a sociology of knowledge can explain a lot and obviate many difficulties. But it cannot pass the ultimate test. It cannot explain how the many different systems knowledge is said to be relative to are related to each other. The failure to pass this ultimate test, rather than any particular doubt about a sociological explanation of knowledge, let alone any more general doubt about relativism as such, obliges us to resist the lure of sociology.

If one believes that any belief or knowledge system, any culture or any society is a self-contained whole which defines its meanings internally through its rules and uses and contains nothing which 'refers' to events or standards or norms outside itself, one is committed to the view that these systems emerge and disappear spontaneously, without rhyme or reason, and are not related to one another, and certainly not related to one another in a way which would explain their emergence and disappearance. Since for millennia it was believed that there is some kind of providence which decrees the appearance and disappearance of these systems, the absence of a special explanation has not been a problem. In very ancient times this law of providence took the form of the theory that the original golden age gave way to a silver age; and, in turn, the silver age, to an iron age. Later, there were theories in which the order of succession went in the opposite direction. Eventually a wider understanding of social structures and economic conditions made way for the theory that these systems followed one another in a succession which started with barbarism, then led to societies in which people were hunters and gatherers. Next followed societies of pastoralists and agricultural people and, finally, there followed the stage of commercialism. When the centre of attention shifted, at the end of the eighteenth century, from economics and social structure to knowledge, it came to be believed that the order of succession went from systems of superstition and magic to a stage of revealed religion to be followed by systems of positive knowledge.

Even a superficial glance will tell us that these series of successive stages correspond more or less to what we know of the history of mankind. Each series represents aspects of that history with which we are familiar. In this sense, such series of successions are more or less true because they correspond broadly to what has happened. As long as there was faith in some kind of providence or its secular parallel, relativism – the belief that each single system in the series was a law to itself and contained all meanings necessary to explain itself and none which might explain other systems – could not arise. Each system's appearance and disappearance were explained as the result of an iron law of progress or regress, as the case may be. However, as soon as one reflects critically on this series, one must see that the theory that it exists has no explanatory power. The series simply recounts what we know to have taken place. But since what has taken place only took place once and was followed by the next step in the series only once, a mere statement of such a series cannot explain why the steps followed one another. The alleged law of development, it turned out, was not a law at all, but a mere succession of accidents. Karl Popper showed that these laws were pseudo-laws and called the theory that there were such laws of development 'historicism'.[41]

A reaction against historicism was already under way before Popper had demonstrated the logical flaw in historicism. It had started with Malinowski's rejection of historicism and with his programme of explaining any one cultural system as a self-contained whole. Nothing, he said, need be explained by a reference to an event outside that system. Malinowski rejected historicism and other more genuinely historical explanations; and conceived an explanatory strategy which came to be known as functionalism. In functionalism, all myths, institutions, habits, thoughts are explained as functionally related to other parts of the same system. In this way reference to the historical past as well as to the synchronic outside world is obviated.

Apart from Malinowksi's own books about the Trobriand Islanders, the great classics about such closed systems, the appearance and disappearance of which would have to remain unexplained and a matter of chance, were Oswald Spengler's *The Decline of the West,* which appeared in Germany soon after the First World War; and Ruth Benedict's *Patterns of Culture,* which appeared in America in 1935. Spengler painted a picture of seven totally unrelated cultural systems which happened to have succeeded one another in time.[42] Ruth Benedict painted a picture of three totally different cultural systems which happened to coexist but which enshrined completely different and incompatible moral and social norms as well as different and incompatible perceptions of cosmic order. The relativism of meanings and the absence of reference of any signifier or code to anything outside the system it was part of were total and final. Neither Malinowski nor Spengler nor Ruth Benedict was able to offer an explanation why these systems had emerged when and where they had. Nor – which is interesting – did they feel that there ought to be an explanation. The closest Spengler came to an explanation was his belief that all systems must disappear just as biological organisms die. The closest explanation Ruth Benedict came to was contained in a proverb of the Digger Indians which she used as a motto. It said that 'in the beginning God gave to every people a cup of clay, and from this cup they drank their life'. In other words, when every thought and institution is relative to a given system and does not refer to anything outside that system, there can be no explanation of how and why systems emerge and how they are related to one another. With her motto, she came as close to creationism and to Linnaeus' belief that every species is an idea in the mind of God as any twentieth-century non-religious person can.

The final chapter in this admission of failure was written by Wittgenstein, Kuhn and Foucault. As stated, such failure was neither an omission nor wilful. It was the logically necessary consequence of relativism. Wittgenstein called the emergence of language-games 'spontaneous'.[43] Not that he could have, on his own premisses, provided any other explanation for their emergence; but, what is more, he saw no need for anything more substantially explanatory. Thomas Kuhn is a little more committed to an explanation of why paradigms disappear and new paradigms arise. He tried to show that the disappearance of paradigms is related to an intellectual unease about the explanatory theories they make possible. Even then, he continues, it is not the unease as such which makes people drop the paradigms, but the fact that the scientists who hold them eventually die.[44] In this way he explained at least the disappearance of any one closed system with the help of gerontology, which must be seen as a branch of biology. Michel Foucault is even more committed to a lack of explanation. In his view, the dominant *episteme* (something remotely similar to a Kuhnian paradigm) changes at least once a century. In the middle ages, the *episteme* prescribed that all explanations were to be in terms of resemblances. In the eighteenth century, explanations had to be in terms of representations; and in the nineteenth century, in terms of time sequences. In our own century, Foucault, following Nietzsche, believes that we explain phenomena as 'masks' of something else. Again, one can see that this succession of *epistemes* does in fact correspond broadly to what actually happened. But like all relativists, Foucault not only is not able to explain why and how these *epistemes* followed one another; but also fails to see that an explanation is called for.[45]

We can see that in some of the most influential thinkers of our century, relativism, whatever its advantages, has failed to pass the ultimate test. It leaves us with spontaneous emergences and random sequences of such spontaneous emergences. Randomness and chance play a very large part in evolutionary explanations. For evolution proceeds by chance mutations and the selective retentions of some of these mutations. The concept of spontaneity is important and valuable. Its explanatory failure in Wittgenstein and the other relativists does not, therefore, consist in the concept itself; but in the failure to associate it with selective retention. The dynamics of evolution requires spontaneity; but that spontaneity is, in turn, controlled by selection. Without selection, it would not explain evolution. Spontaneity in isolation is not an explanation, but a failed explanation. Selection is only conceivable if and when something in the spontaneously emerged organism or system refers to something outside the system, i.e. to something that does the selecting or according to which the selecting is done.

In spite of the inadequacies and shortcomings, relativism is enjoying much popularity because it claims to obviate the need for genuine evolutionary explanations. All known systems, cognitive paradigms or cultures, are a departure from the original conditions of our distant hominid ancestors. There have been changes. If one rejects relativism as the last word and is looking for the reasons for these changes in order to explain how these systems are related to one another, one must come up with a taxonomy which is hierarchical. In such a taxonomy, some systems will be further away from the primitive starting point than others, and they will not all be equidistant from that point. In such a taxonomy there will be an explanation of why the social systems of masses of people which populate Manhattan have developed later than the feudal agrarian systems of medieval Europe. Equally, there will be an explanation of why the Theory of Relativity came after Newton and why quantum mechanics came after classical physics and why this order could not have been reversed. In other words, in such an explanation one will see that the relationship between the passage of time and the changes which have taken place is not random. But if one takes relativism as the last word, such changes as have taken place must be seen to have happened by chance. One will then be left with a taxonomy which is similar to the taxonomy of Linnaeus and in which all systems and paradigms are equidistant from the point of departure. For relativists, the changes which have taken place are randomly related to the passage of time and could have taken place at any time and in any sequence whatever.

It is clear that these two taxonomies have political implications. The first, non-relativistic taxonomy is hierarchical. Some systems are further removed from the starting point than others. In the second taxonomy all systems are equidistant from the starting point. In our age of humanitarianism and egalitarianism, the second taxonomy is more likely to be in harmony with the social and political aspirations of *all* human beings than the first taxonomy. If one can persuade oneself that shamans and medicine men are no different from the physicists of the Cavendish Laboratory and that there is no difference between their cognitive claims and that it was a matter of chance that the Cavendish physicists emerged after the shamans and the medicine men, rather than before, one can approach the multitudinous differences of all the social and cultural systems that present themselves at the United Nations Headquarters in a genuine spirit of cooperation. Obviously, a taxonomy in which all systems are equidistant from the point of departure is politically more appealing than a taxonomy in which they are not. The notion of equidistance makes for tolerance and is therefore morally superior, even though it fails to explain how the many systems are related to each other. This conclusion is based on a politico-moral consideration and cannot be taken to settle the dispute between relativism and non-relativism. If one took it so, one would have to presuppose that the substitution of sociological factors for cognitive factors in the decision-making process is legitimate. Nevertheless, this conclusion goes a long way towards an explanation of why, in the second half of the twentieth century, relativism and the sociology of knowledge on which it is based are so popular and why to so many well-meaning and intelligent people their lure is proving so irresistible.

4

THE NATURE OF THE MIRROR

Knowledge is a relationship between two terms, at least one of which contains information about the other. We have not yet been able to find an answer to the two great questions we have asked – where does the information get stored, and how does it get from one term to the other? In a search for an answer, we have come along several different roads only to find that, contrary to proverbial expectations, none has led to Rome. When we tried to come from within and started with consciousness, we found that consciousness does not really store information and that such information as it does store cannot be represented. What is more, we found that what consciousness does succeed in representing is not information about the world. We fared differently but no better when we tried to come from the front. The idea that information is transferred when or because there is a causal link between the two terms turned out to be a pipe-dream because such causally induced information cannot stand on its own two feet but is in need of linguistic expression. Thus the information gets bushed, and by the time it finally comes out, it is no longer causally related to the first term of the relationship.

Next, we tried to find out what we get when we take it that the information comes from behind. The possibility that the information is actually generated by the rules of sociability, including the rules of language, seemed more promising. We were able to show that the principles of the invariance of experience and of the need for consistency of explanations can be derived from the conditions of sociability. It was only in the very last instance, when we were looking for an explanation of the fact that all actual sociabilities are related to one another, that the recourse to sociability failed. The view that knowledge is a social convention must enshrine relativism. It is quite viable, until it fails to explain how one relative system is related to another relative system. Such an explanation is indispensable because we know that all systems are part and parcel of one and the same universe and must therefore be compatible with one another, or – to put it negatively – cannot be left to agree to differ from each other. This kind of explanation of the ultimate compatibility and consistency of all conceivable systems, it seemed, is only possible if one is able to consider each system as a function of something other than itself.

I

The one dimension we have not yet explored is the past. The past as the source of information has a long and very venerable history, even though it is only during the last century that any real progress in the understanding of the role of the past in the transfer and storage of information has been made. It can be no surprise that the recourse to the past should have such a long history. Imagine the possibility – if that is at all possible – of an organism without a past suddenly erupting into or appearing in this world. Not even science fiction stories in which an alien suddenly arrives entertain this possibility – for even an alien comes from somewhere, where he or she is *not* an alien. We can only think of such an appearance as an epiphany, for any non-miraculous appearance would have to be the result of prior events, that is, the organism in question would have to have a past. But let us suppose that such an epiphany had taken place. The organism in question could not survive, because it would starve to death as well as suffocate long before it would have time to pick up the necessary information as to what to eat and how to breathe, even if such picking up and such learning from the instruction offered by the environment were at all possible. The whole notion of such a *tabula rasa* organism is so patently absurd that not even the great propounders of strict empiricism were able to sustain it completely. Locke, for example, believed that we are born, though without innate ideas, with 'natural faculties',[1] and Hume, a little later, took it tacitly for granted that certain impressions are given in an objective and regular succession, i.e. that we are capable of recognising that they are so given.[2]

Contrary to Locke's and Hume's professions, but consistent with their own assumptions, we all, including new-born babies, know much more about the world than our observation of the world would warrant. Locke's trouble and the trouble with his countless imitators right down to Skinner in our own century was that they thought that the past could be disregarded as long as one had satisfied oneself that there are no innate ideas. It never occurred to any of them that the influence of the past is far more widespread than the generation of innate ideas. We come into the world with a lot of information, even though it may not be in the form of ideas, and this fact points clearly to the past. None of us ever behaves as if he or she were new to the world and had no ancestors and no past, and since even the most famous *tabula rasa* philosophers admitted tacitly that the past does have a formative role, the past must be a promising candidate for an answer to our questions as to where the information comes from and where it is lodged.

It has been known for a very long time – at least since the first book of the Old Testament was put together – that living matter and the rest of the world fit together and that living matter fits into the world and contains an awful lot of information about the world. No matter which of the many possible definitions of living matter is used, living organisms contain an enormous amount of information about the world as well as about each other. The presence of living matter, no matter how incomprehensible its functioning and its origin are, cannot be a miracle, for it fits the rest of the world. In the first instance, therefore, living matter is the place where information is stored. That much was clear even to the people who wrote the text of the first book of Moses. What was less clear, and what has been debated ever since, is how the information got there. In very early days, when people gave the question some thought, they came to a very obvious conclusion and formed the theory that since the relation between living matter and the rest of the world was such a good fit, it must have been so designed by a Designer. Ever since the Enlightenment, this theory has come in for a lot of ridicule. In view of the fact that there is no independent evidence for the presence of a Designer, it is indeed a theory which is not very plausible. But one should not scoff too readily, for it is quite an ingenious theory which bears clear witness to the recognition that there is fitness and that the presence of living matter is not an unpredictable accident and that we are not aliens from outer space. Even if one remains sceptical as to Richard Dawkins' explanations of the fit, the importance of the massive evidence of fits he has piled up in *The Blind Watchmaker*[3] cannot be exaggerated. The argument from design only seems ridiculous because it has been taken to mean that there was, in the first place, before the world existed, a Designer. But if one looks at it the other way round – that the world is a place where everything, or almost everything, fits together – the theory that it must have been designed is no more absurd than many another theory. It is simply an attempt to explain a number of facts which are there for everybody to see.

The next great step forward in the attempt to explain how information was and is transferred was taken by Plato. Plato focused on one single aspect of such information transfer, to be precise, the aspect which we conventionally label 'knowledge'. The authors of the first book of Moses, acknowledging that the all-pervasive fitness is neither an accident nor a miracle, had looked to the past. In the past, they said, there must have been a Designer. Plato too acknowledged the fact that our knowledge of what goes on in the world cannot be an accident or a miracle and that the presence of knowledge must have something to do with the past of the people who have that knowledge. Ignorant of biology in general and genetics in particular, it was not easy for him to think of a theory that would explain the fact that people have knowledge when they come into the world, that is, *before* they have been exposed to it. But if he was ignorant of biology, he was well versed in Orphic mythology and the doctrine of the transmigration of souls. He thus came up with the ingenious theory that we recognise chairs as chairs and horses as horses, no matter how particularly different one chair and one horse are from other chairs and other horses, because before our souls were entombed in the body, they had pre-existed in a realm in which they had been able to 'see' the idea or form of a chair and a horse. The ability to recognise a chair as a chair was therefore basically an act of recognition which depended on the prior existence of the soul, that is, on its past history.

The great difficulty with this daring hypothesis was first, that there is no evidence other than the ability of recognition for the pre-existence of the soul in another, spiritual realm. Second, given the vast variety of objects and events to be recognised, it is statistically quite incomprehensible how a soul could have visually stored up that much information about the ideas or forms of all these countless objects and the mutations and transformations they can be subject to and which they can undergo. Plato's theory, though it remained popular for many centuries for want of a better one, was not satisfactory because of its manifest reliance on mythology and its lack of formalism. Nevertheless, Plato had hit upon the right problem and was looking for a solution in the right direction: we come into the world with a lot of knowledge and we must suppose that that knowledge – whatever its precise shape – was acquired in the past, i.e. before we came into the world.

The next great step forward was taken two thousand years later by Immanuel Kant. Kant held fast to the idea that an initial ingredient in all knowledge must come from the past because it would be impossible for an organism, no matter how clever, to arrive at, say, Newton's laws of gravity, if it had nothing but a *tabula rasa* for a mind. He also realised, looking at Plato, that one could not reasonably suppose that the soul had, before its encasement in a body, stored up information about all possible individual objects and the events they might be enmeshed with. He reasoned instead that the human organism must be in possession not of the myriads of individual objects it knows or can know of, but of the formal categories by which they are recognised. In one fell swoop, he made Platonism statistically plausible by the argument that the a priori knowledge we have is knowledge of the formal categories which all conceivable objects share.

Kant, needless to say, also rejected Plato's mythology about the pre-existence of souls and suggested instead that there are compelling rational grounds why all human beings use the categories and the forms of perception he had listed and why they cannot use others. In part, this argument was his transcendental deduction of the categories. Philosophers have worried about the validity of this deduction ever since, and it has been shown, often very convincingly, that the reasoning used by Kant is not absolutely compelling and that one could also arrive at other categories by similar reasoning. Worse, the discovery of non-Euclidean geometry and the appearance of General Relativity as well as of quantum mechanics have shown that Kant's forms of perception (time and space) are not always applicable and, therefore, not necessary and that his insistence that we cannot perceive anything other than by locating it in Newton's space and time is misplaced. Most amazing of all is the deep irony in Kant's procedure. He had set out to assign limits to pure reason. But in the event it turned out that his critical philosophy was insufficiently critical of the power of reason to carry out the transcendental deduction as well as of the power of reason to determine that absolute space and time are the necessary forms of all perception.

Let us be more precise. Kant carried out the transcendental deduction, and the critics who have since tried to improve on his deduction[4] are hardly more convincing than Kant. The real limitation of the deduction does not lie in the reasoning employed, but in the conception of the deduction. The deduction leads to the establishment of those categories that have to be in place before any experience can be registered and indicates the conditions of experience which cannot be gathered by experience and are not learnt through instruction from the world. But the deduction is carried out in relation to one particular set of knowledge of the world, to be precise, in relation to Newton's system. It is not carried out in relation to nature itself and does therefore only indicate those categories that have to be in place before Newton's physics can become established and credible. It does not indicate the categories that have to be in place before we can know anything about raw nature. Hence, the transcendental deduction established, by the way it is defined, the dichotomy between noumena and phenomena. It places all knowledge of the former beyond our ken; and confines all knowledge to the phenomena – to Newtonian physics, in this case. This limitation of the deduction of the categories to a particular perceived world, rather than its application to the world as such, makes irrelevant the question whether the deduction itself could be improved on.

Kant thus drove a wedge between the world as it is in itself and the world as we perceive it. His conclusion that we 'prescribe' the laws of nature to the world follows logically and is a truism, rather than a startling new orientation of our thinking. Bertrand Russell put it very bluntly[5] when he said that when we are wearing blue spectacles, the world will look blue. It can be no argument in favour of wearing blue spectacles to say that we are wearing blue spectacles, because the world *appears* to be blue. In looking towards the past, that is, in looking at the conditions which temporally have to precede all knowledge, Kant had failed to grasp that those conditions develop or evolve or emerge by interplay with the world and, if they do, that they refer to the world as it is in itself and that the dichotomy between noumena and phenomena cannot be absolute.

Some nine years after the publication of the *Critique of Pure Reason* Kant considered whether these conditions might seem to stand, after all, in a more really necessary relation to the world. He wrote that

> there has to be in the subject a ground which makes it possible for our representations to originate the way they do and not in a different way in which they refer to objects which are not yet given. That ground is innate.[6]

Somehow, Kant seems to have been working forward to the view that these conditions are fundamental and not just the preconditions for knowledge of the Newtonian world. But this is as far as he got. The possibility that these conditions are the result of evolution and come from natural selection could not possibly have occurred to him. Kant was a good evolutionist where the planets and the solar system was concerned. But in his time, there was no way in which he could have applied the concept of evolution to the human mind and its nervous system and recognised that the world and the way we perceive the world are intimately related.[7] Had he been able to progress towards an evolutionary view of perception and knowledge, his Copernican Revolution would have appeared in an even more startling light. As it was, it would have been more correct for him to speak of a Copernican Counter-revolution; for his way of reasoning had put man right back into the centre of knowledge: the world we know, Kant had argued, is not the world as it is in itself, but as it appears to man, and the only reason why it cannot appear differently is that there is a logical necessity in the very moulds in which it appears to man, and all experience by which we test its characteristics and features must go through those same moulds and cannot therefore but confirm the way it appears. The world in itself, so to speak, has no standing in our knowledge and cannot make itself felt or heard or seen or sensed. But had Kant been able to work forward to the view that our moulds of perception have evolved through natural selection and are therefore an integral part of the world we perceive, the man his Copernican Counter-revolution would have put back into the centre would not have been Ptolemy's man, but a man who was an integral part of the perceived and known world and who, because of his standing as an integral part of that world, had direct access to the world as it is in itself.

II

The next and decisive step forward was taken by two almost casual remarks Darwin jotted down in his notebooks. 'Plato says in *Phaedo* that our "*imaginary ideas*" arise from the preexistence of the soul, are not derivable from experience. Read monkeys for preexistence.'[8] On another occasion he remarked that 'he who understands baboon would do more toward metaphysics than Locke'.[9] One notes that with these two statements, Darwin wiped out the two main opposing strands in the whole of western philosophy – the idealist as well as the empiricist tradition. But one must also note that he did not do so evenly. He obviously saw some merit in Plato and suggested that Platonic thinking be amended. The recourse to the past which he had detected in Plato was obviously correct. Locke, on the other hand, was dismissed out of hand, because the notion that man is a *tabula rasa* and that everything we know has been learnt by watching the world is patently absurd. Both remarks, taken together, amount to the assertion that there is a lot of a priori knowledge and that that knowledge is lodged in every organism because every organism is descended from ancestors and, therefore, has a past. Although it has since become clear that Darwin himself took a systematic and conscientious interest in philosophy[10] and that these fertile remarks must, therefore, not be seen as either fortuitous or incidental, they fell, unfortunately, on deaf ears. Philosophers blithely continued to wonder how man, from the Olympian position he occupied as if he were an alien in the world or a mere spectator or as if he had been placed there by God, as it were, as an afterthought, might be able to understand the world. They came up with a number of suggestions, none of which have proved very compelling, so that Einstein considered such understanding nothing less than a miracle. At a time when Darwin's thinking had revolutionised just about every single science in the world, Ludwig Wittgenstein declared portentously in his *Tractatus* that the Darwinian theory has nothing to do with philosophy.[11]

The threads were finally picked up and put together by a biologist. In the early years of the Second World War when, by a strange coincidence, he was a professor in Königsberg, Konrad Lorenz used Darwin's idea about the formative role of the past to put the finishing touches to Kant.[12] He argued that Kant's scepticism about what the world is really like was unjustified because the cognitive structure which enables us to know what the world is like had evolved through natural selection. The reason why our minds have this particular and no other cognitive structure, Lorenz pointed out, must be that we have evolved and have not flown in, so to speak, from outer space. Our cognitive structure has been selected by and, therefore, reflects or represents, the real world. The categories of Kant, or something comparable, are not accidental. They are not created arbitrarily and could not be different. They are adaptive. Whatever they tell us must be some kind of truth about the world and the environment as it really is. Once the evolution of the cognitive structures is taken into account, the Kantian distinction between the world as it in itself and as it appears to us becomes unnecessary. Unbeknown to Lorenz, Hegel had already used an unbiological but formally very similar argument against Kant's absolute dichotomy of noumenon and phenomenon.[13]

The third foundation for a proper appreciation of the fact that most of our information about the world comes from the past was provided by Karl Popper in 1934, when his *Logik derForschung* was first published. In this book Popper provided the first philosophically tenable account of information transfer or, as he then called it, of the acquisition and growth of knowledge. At that time he was exclusively concerned with the discovery of a criterion of demarcation between science and metaphysics, or knowledge and non-knowledge. He pointed out that the criterion of verification, which was and still is widely used, cannot serve this purpose because the logical structure of a general proposition (of theory) is such that no amount of limited verifications or instantiations could prove that a general proposition was true. By contrast, he showed that it is possible to falsify a general proposition conclusively, because one single falsifying instance would, theoretically, suffice.

Hence he argued that one can tell the difference between a statement about the real world and a statement which merely and speculatively suggests that all gods are invisible or that all souls are immortal by attempts to falsify them. If the conditions under which it would be falsifiable can be indicated, the statement refers to the real world. If no such conditions can be indicated, the statement refers to nothing at all.

Though the logic of this argument is impeccable, it raises a number of new problems, most of which have never been completely solved. To begin with, even when falsified, any theory can be salvaged by *ad hoc* hypotheses. The question then arises as to how many such *ad hoc* hypotheses are legitimate. Equally thorny is the consideration that an attempt at falsification, the conceptual terms of which are themselves dependent on the theory to be falsified, is not likely to succeed; whereas an attempt at falsification, the terms of which are dependent on a different theory, is almost bound to succeed, so that the successful falsification could reasonably be disregarded. In short, the theory of falsification, while solving some of the oldest problems in philosophy, opened up a number of new problems. Nevertheless, it was of epoch-making importance because of its hidden agenda.

The hidden agenda concerned its implications for our understanding of information transfer and were finally brought out by Popper himself many years later, in his *Objective Knowledge* of 1972. We learn and pick up information, the theory states, by selecting it from an abundance of conjectures which are guesses and stabs in the dark. In this form, the theory is directly opposed to the view that we pick up information by observing the world and by being, so to speak, instructed by it. The 'knower' makes proposals, and these proposals are scrutinised by reality. The process is, of course, not automatic. The scrutiny is the result of conscious criticism carried out by the proposer or by other people in the light of their experience of the world. The process of information transfer is therefore a special form of falsification. First, a vast array of theories are proposed, and then, all but those which fail to be falsified are rejected. In this way, the process of information transfer can be seen to be continuous with organic evolution itself. In evolution, there are chance mutations and a selective retention of those that fit better than others. In the conscious acquisition and growth of knowledge, there are chance or almost random proposals; and retention of those which have withstood attempts at falsification.

The most important and crucial contribution of *Objective Knowledge* was this extension of this theory of information transfer to the pre-human, animal world. While one could and had imagined that human beings learn by observation and by taking instruction from the world (a method commonly known as induction), one cannot by any stretch of the imagination suppose that animals can pick up information in the same way. The enormous merit of Popper's theory is that it is equally applicable to animal and human learning and that it abolishes the gap which the conventional wisdom of philosophers had insisted exists between pre-human and human information gathering. The abolition of the gap follows from Popper's argument that the acquisition of knowledge comes in all cases from the selection from an abundance of proposals, rather than from instruction by a limited number of given facts, or sensa or data for short.

The philosophy of Popper provides a good starting point for *philosophical* Darwinism, which should not be confused with what has become known as evolutionary epistemology.[14] In suggesting that we take falsification as the hallmark of *empirical* content, he rejected the idea that we acquire knowledge by following the instructions we receive through our senses and thus insisted that knowledge *precedes* experience. He did not explicitly say that in preceding experience and in being a priori, it comes from the past. But he did say that it arises earlier than experience and in this way rejected any kind of philosophical Lamarckism.[15] Such knowledge is hypothetical and cannot be distinguished from fantasy or mistaken knowledge any more than in Darwinian evolution the emergence of monsters and ill-fitting organisms can be prevented. What can and is prevented, however, is their survival. Similarly, in Popper's view, false or hallucinatory knowledge can be distinguished from genuine knowledge by attempts at falsification – attempts which can begin *after* the proposals have been made. Survival after such attempts at falsification have been made makes the knowledge provisionally or hypothetically true. Thus truth is approximated by the selective retention of a priori hypotheses. This model of the acquisition of knowledge is thoroughly Darwinian and, in fact, goes a long way towards the establishment of philosophical Darwinism. Unfortunately, in his earlier writings, Popper made a number of non-Darwinian and even anti-Darwinian asides because he mistook Darwinism for some kind of developmental law (historicism)[16] and thought of Darwinism unfelicitously as a 'metaphysical Research Programme'.[17] Although in his *Objective Knowledge*[18] there is a full retraction and apology for these 'contemptuous' remarks, there are many people who think of him as an opponent of Darwinism, and he has even been cited in support of creationism in the notorious court action against evolution in Tennessee.[19]

III

In factories, workshops and commercial artists' studios, it has been known for a long time that the most efficient and effective way of transferring a letter from one piece of paper to another is to make a stencil, place it above the other sheet of paper and then apply an abundance of uncontrolled colouring to the stencil. The shape of the letter, cut out of the top sheet, will then appear accurately on the bottom sheet. The method of transfer consists of the uncontrolled application of an abundance of colour to the top sheet and of a selective retention of colour by the bottom sheet. The other method, which consists in an effort to copy the letter minutely, by following its contours, is not only laborious, but also unlikely to result in an accurate replica. In contrast to practical people in workshops and factories, philosophers and scientists have been reluctant to adopt this method of information transfer as rational, let alone as reasonable. Indeed, if one has no practical experience, the other, laborious method must obviously present itself as the rational method because it is based on following instructions. It also seems more rational, because it seems more economical to sharpen one's powers of observation and follow instructions. That way, it is believed, one avoids mistakes and waste. However, such economy is more apparent than real. For in following instructions, no matter how minutely, one will not be able to carry them out minutely enough to avoid inaccuracies. The time saved, therefore, is being wasted at the other end. By reason of this seeming, if unreal, economy, the thought that selection from superabundance rather than instruction by a limited number of observations is the key to information transfer has been slow to establish itself.

Wherever we look, we find that the idea that learning takes place when instruction is being followed has appeared as the initially plausible, first deliverance of wisdom. Lamarck thought of evolution as a process in which instruction is being followed and the registered results passed on to the next generation. In immunology, it was believed that the entry of foreign bodies instructs the organism to produce the appropriate response, until Jerne demonstrated that it is more likely that a process of abundant production of antibodies precedes the entry of the foreign body and that the entering foreign body then selects from that abundance those anti-bodies most hostile to it. Antigens are not, as instruction theories had thought, templates, but selective carriers of spontaneously circulating antibodies to cells which can reproduce them. 'An animal cannot be stimulated to make specific antibodies, unless it has already made antibodies of this specificity before the antigen arrives. It can thus be concluded that antibody formation is a selective process and that instructive theories of antibody formation are wrong.'[20]

Or look at neuroscience. It was long held that neurons grow along certain paths because they have been instructed to do so by genes. But now the work of G. M. Edelman is based on the idea that neurons appear in rich abundance and that the environment in which the organism is living slowly selects a limited number from that abundance:

> the organism receives stimuli from its environment or econiche as polymorphous sets. As a result of action, stimuli select among various dynamic nervous system states and arrangements that have already been established prior to the receipt of these stimuli, leading to enhancement of some states and suppression of others. Such stimulus sets constitute information, in the instructionist sense, only *after* selection, response, and memory have occurred; and information processing, in the larger and more specific sense of the term, occurs only *after* social transmission has emerged as an evolutionary development.[21]

Similarly, Popper's contention that we learn by error elimination, that is, by retentive selection from a superabundance of proposals, is a rejection of the impractical method of learning and acquiring knowledge by committing specific instructions to memory. Selectionism is more realistic, more practical and more sophisticated than instructionism.[22]

However, we must guard against the temptation of using the historicist argument that selectionism is better because it is later than instructionism. If we used such an argument, we would be guilty of bolstering selection in the same way in which the positivists of Chapter 2 tried to bolster their faith in the beneficial effects of observation. The difference is that positivism, as we have seen, has nothing else to commend itself and was therefore wholly dependent on the historicist argument that it was the latest and most up-to-date model; whereas selectionism can very well stand on its own two feet and does not require the historicist recommendation that it is the latest fashion and up to date. It just so happens that it is. There is, however, a possible explanation why instructionism came first and why it has proved so tenacious. The conception of instruction is modelled on the push-and-pull causality of classical mechanics and as such carried the hallmark of sound and observable good sense.[23] It was almost to be expected that with the growth of physics in the seventeenth century there should go, hand in hand, an epistemology which said that knowledge is generated by pushes, exercised by the world on the mind; and that nature thus instructs the knower. There was no viable alternative epistemology until biology started to study living matter and showed that information transfer does not take place by instruction, but by selection from an abundance of proposals. Here we have a case of a sociology of knowledge in reverse. Whereas sociologists of knowledge are fond of explaining the emergence of classical physics as the result of ideological pressures,[24] we can here surmise that we have an example in which philosophical ideas were shaped by the success of physical theory. Baconian positivism, to be sure, preceded both Descartes and Newton; but had it not been for Descartes's and Newton's successes, Bacon – not to mention Locke! – would not have worn so well.

The idea that we learn by following instructions issued by the world continued its hold also for a different reason. It was considered that instructionism alone guarantees that there will be a correspondence between knowledge and the world. No other method of information transfer seemed to include a similar guarantee. But this consideration is without foundation. Instruction by the world cannot lead to correspondence, because there is an inevitable hiatus between the impact the world makes and the articulated statement or representation of that impact. It is simply not the case that the world is prestructured into a set of 'facts' and that truth by correspondence consists in language which portrays these precategorised 'facts'.[25] Protocol sentences, and the newly revamped neuroscientific version of Protocol sentences according to which labels attach themselves magically to neuronal events, are, as we have seen in Chapter 1, chimeras. Correspondence, on the contrary, can only happen if the relation between knower and known is not the result of instruction. This consideration makes a prima facie case for some kind of selection as the only viable method of information transfer. The most instruction can ever be imagined to lead to – because of the hiatus between sensory impact and articulation – is a coherence of verbal hypotheses about impacts, i.e. the very opposite of what has been expected from instruction. Against all expectation, instruction does not lead to correspondence, but, when hypothetically verbalised, yields the sort of coherence one is entitled to expect from intelligent speech. This articulated and verbalised instruction, as was shown in Chapter 3, can be coherent with other messages similarly produced by following instruction, even though or, perhaps because, it cannot be seen or proved to correspond to the causes of the impact. But then, it is not coherence of statements, but correspondence with the world we are after.

IV

We now have three general contentions: (1) Darwin observed that every organism has a priori knowledge because it has a past. (2) Lorenz observed that it must be the noumenal world, not a phenomenal world, which manifests itself in that knowledge. For the organism which has knowledge is itself part of the noumenal world and not an outside observer who occupies some Olympian stance. Knowledge is therefore to be seen as a form of self-reference of the real world to a part of itself. The knower, so to speak, has been shaped by and out of what is known and must reflect, or represent and even mirror, at least up to a point, what it has been shaped by. (3) Popper observed that the process of such shaping and of such adaptation is continuous from 'the amoeba to Einstein', that is, that knowledge is acquired, both by organisms and by the mind of *homo sapiens*, by chance mutation and selective retention and not by taking instruction. The neat formula 'by chance mutation and selective retention' is due to Donald T. Campbell's paper 'Evolutionary Epistemology' of 1974.[26]

On the basis of these contentions we can now formulate the philosophical consequences of our biological knowledge. I will call them the central contentions and hope they will not be broadly confused with the study of what has become known during the last two decades as evolutionary epistemology. In my view, these central contentions are more specific than anything that has been claimed for evolutionary epistemology and are as a coherent body more vulnerable to the several criticisms which have been levelled against the many diverse and frequently inconsistent views which go under the name of evolutionary epistemology. A detailed discussion of these criticisms has to be left to the next chapter.

Building on both Popper and Lorenz, the first attempts at an evolutionary epistemology were due to Gerhard Vollmer's *Evolutionäre Erkenntnistheorie* of 1976[27] and soon after in 1979 to Rupert Riedl's *Biologie der Erkenntnis.*[28] Since then there have been several anthologies on the subject.[29] They cover a vast field of very diverse claims and arguments of all sorts of matters either vaguely or directly related to biology and range from ethical determinism to sociobiology, from cognitive science to ethology, and touch only rarely on philosophy. It is often taken that evolutionary epistemology is concerned with the attempt to show that every single element in the human cognitive apparatus is inherited from man's ancestors and that the appearance in man of consciousness is an inexplicable fulguration which does not have any clear adaptive advantage and which may have to be considered an inhibiting and possibly even a destructive factor. In this sense, no more is meant by evolutionary epistemology than the study of the natural faculty of cognition; and often such cognition is taken to be no more than a faculty capable of taking instruction from the world and representing it. In this sense, evolutionary epistemology is considered synonymous with naturalised epistemology – another topic which will be discussed in detail in Chapter 5.[30] Apart from including too many diverse topics, evolutionary epistemology also keeps steering clear of a broader philosophical approach.

Indeed, in the wake of Konrad Lorenz, evolutionary epistemology has become a palliative to positivism by its ability to demonstrate that the mechanisms of perception are due to natural selection and can be taken to mediate real knowledge of the real world. It provides good biological proof of why and how organisms could be seen to be learning from instruction. Where Darwin had suggested we discard Locke and amend Plato, the followers of Lorenz seem to be discarding Plato and to be amending Locke. In any case, they cling to the belief that while mechanisms of perception are due to evolution by natural selection, knowledge itself is founded upon the inductive accumulation of the particular bits of observations, the veridical nature of which is guaranteed by the fact that the mechanisms of perception are fool-proof because naturally selected to be so.[31]

As long as attention is confined to those faculties capable of registering information and to their evolution, one is bound to miss out a crucial aspect. For evolution to take place, there has to be an abundance of proposals from which to select. Thus the making of mistakes in reproduction is essential for evolution. In organisms it has been discovered that there is enough room for error in the reproduction of DNA to allow for those necessary mistakes. When we are dealing with knowledge which, though evolving, is not evolving by the reproduction of genes, we have to look for a different source of error. Therefore, in order to understand the evolution of knowledge, we must not confine our attention to the evolution of those faculties which find information; but to those faculties which fail to find information and the presence of which, though a positive advantage, must appear, to cognitive scientists who are studying the adaptiveness of organs and faculties, a direct hindrance. As a result, the pursuit of evolutionary epistemology has limited itself to the study of organisms, their adaptiveness and the faculties which make that adaptiveness possible. With the exception of Karl Popper and Donald Campbell, evolutionary epistemologists had to confine their study to those organisms in which the occurrence of the essential errors could be attributed to the behaviour of DNA and had to keep away from a consideration of the knowledge the evolution of which depended on a more devious strategy of making mistakes. In any case, we will locate the faculty for making mistakes in the phenomenon of consciousness and its oblique by-product, three-dimensional language. Neither consciousness nor three-dimensional language is a faculty, the evolution of which one can study together with the evolution of, say, the eye or the tactile neurons on the surface of skin. On the contrary, as long as consciousness was studied by psychology or cognitive science as an adaptive faculty, it turned out to be curiouser and curiouser, as Alice in Wonderland correctly observed. Only philosophical reflection is capable of understanding its oblique function of providing the abundance of mistakes from which selection can be made.

Evolutionary epistemology, so far, has handicapped itself by making a naturalistic study of the evolution of those organs which can get it right. Paradoxically, we need a bit of non-naturalistic philosophy in order to understand the process of natural knowledge acquisition. For this reason, it is preferable to abandon the term evolutionary epistemology altogether when we are embarking on a study of the role of the faculty which can get it wrong, that is, of the faculty which, in getting things wrong, provides the raw material for evolution. Recombinant DNA suffices only to produce an abundance of organic proposals. In order to explain the evolution of knowledge which is not embodied in organisms, we require a special faculty for making faulty proposals over and above the mistakes made by DNA.

The first contention is that every single organism comes into the world with a lot of information about the world. Each living cell, prokaryotic as well as eukaryotic, is surrounded by a membrane which enables the cell to maintain an internal molecular composition which is completely different from that of the medium in which it lives. This membrane is selective in the types of molecules which it allows to enter and to leave the cell. The membrane, therefore, must be seen as knowing what it will allow in and out. The cell, surrounded by its membrane which divides the cell from the rest of the world, is the model of knowledge. The cognitive relationship is a relationship between what is on one side of the membrane and what is on the other side of the membrane. The cell has a lot of information about the world it is living in. This information is present a priori – that is, it is not packed in because the cell has been exposed to the world and experienced it, absorbed it and registered it. The medium in which the cell is living does not instruct the cell as to what the membrane should let pass, and the cell's knowledge of the medium it is living in is not learnt by observation of the medium in which it is living. The membrane, in dividing the cell from the rest of the world, establishes a cognitive relationship between what is on one side of the membrane and what is on the other. When cells become cooperative and form organisms in which they carry out specialised functions, the other cells become part of the outside world for the single cell. In multicellular organisms, the individual cells are therefore dependent on each other and have become structurally and functionally differentiated from each other to perform various specialised roles necessary for the whole organism. Strictly speaking, therefore, one should now consider the outer skin of the organism as the analogue of the cell's membrane. But since we are here not concerned with biology, but with the philosophical consequences of biology, we can leave this complicating aspect of the matter alone.[32]

The second central contention is that more sophisticated knowledge, the sort of knowledge we hold consciously and express linguistically and associate with science, is equally a priori and, for this reason, continuous with the information transfer and information storage we find in the model of knowledge itself, the single cell. Unlike the knowledge stored in the cell, this knowledge is not genetically programmed and inherited; but it is equally a priori because it is generated before the organism which holds it is exposed to and has learnt from the world. More precisely, it consists of a set of proposals which the human organism makes to the world. How this is done, how they are expressed and how one distinguishes silly ones from less silly ones and good ones from better ones will be explained later.

Let me enlarge on the first contention. The information stored in all cells and all living organisms consisting of cells must be less specific than the information envisaged by Plato; but more substantial and less formal than the information postulated by Kant. Plato and Kant were not only mistaken as to the derivation of a priori information; but also as to the precise content of such information. Though it is a priori as Plato and Kant had claimed, it is in biological organisms stored in the molecular structure of the cells and, therefore, unworded and not capable of linguistic description. To be more precise, linguistic description is possessed by human beings theorising about the molecular structure, but not by the organisms having it. According to the present state of neurological knowledge, there is no way in which we can say exactly how substantial or how formal the information which is available a priori is. But it has been discovered that there is, immediately after birth and during the first months of exposure to the environment, a fine tuning of whatever is stored in the nervous system. Genetically, the embryo is programmed to produce a wild abundance of nerve endings in special directions; more than the organism can do with and more than it needs for survival. Once this abundance of endings is exposed to the environment, the environment, through its physical impacts upon these nerve endings, selects some and allows most of the others to atrophy. In this way the environment itself provides the finishing touches to the information which is present a priori and makes sure that, though a priori, it is adequate to the world. It is important to understand this fine tuning, which assures adequacy of reference, as a continuation of the process of natural selection from an abundance of possibilities; and not to mistake it as a form of instruction of the nervous system by the environment. However, this time it is nerve endings and synapses rather than organisms that are being selected. One might even be tempted to think that neuroscience has now provided an explanation of how something like Kantian categories have been put into place and why they actually do refer to the real world.

> Interactions with the environment contribute to the formation of more and more complex neural organization, despite the meagre evolution of the genetic inheritance. Each generation renews this selective shaping of the brain by the environment. It is accomplished very rapidly compared with the geological time scale of the genome's evolution. Epigenesis by selective stabilization saves time. The Darwinism of synapses replaces the Darwinism of genes.[33]

We become able to categorise events, writes G. E. Edelman, by a 'selective system in which a large preexisting set of variants of neural networks formed during embryonic life … are selected for and against during the worldly life of the animal'.[34] After birth, the movements of the organism bring the organism into direct contact with the shapes of the world. The development of the organism depends on such individual learning. But such individual learning is not a learning by receiving and registering and memorising instructions. It is made possible because the contacts with the world *selectively* strengthen connections within neuronal groups in accordance with the individual's experiences.

The stored information which permits the fine tuning which adjusts the organism to the world is inherited and transmitted genetically. At birth, it is therefore a priori and precedes any experience, but it has been established phylogenetically among the ancestors of the organism, and we are obliged to suppose that that information has been established in the ancestry by natural selection. This means that among the ancestors there must have been countless individuals with less fitting information about the world and that these ancestors had fewer offspring or did not survive to have any offspring at all. At no stage among the ancestors was there an accumulation of information a posteriori, that is, by induction and storage of particular bits of information. It is therefore misleading to say with Konrad Lorenz that this information is ontogenetically a priori, but phylogenetically a posteriori. Misleading, but not entirely wrong. The distinction between a priori and a posteriori was supposed to be rigid. Some knowledge, like innate ideas or categories, was supposed to be a priori; and other knowledge, a posteriori. But as soon as one asks why there are innate ideas or categories and why the innate ideas and categories are the ideas and categories they are and why they are not different, one is led to the realisation that they must have emerged in response or with a view to the world they are about. And thus one comes to the conclusion that they are not all that a priori – and certainly not in any absolute sense of 'a priori'. Nevertheless, it is useful to retain the term, because it indicates that organisms confronting the world are not empty buckets into which information is poured as a result of observations. We should make a change in the meaning of the term. When we call knowledge a priori, we should mean knowledge that is generated – though not validated – before the organism's exposure to the world. But even this meaning is not entirely correct. For organisms which have survived and reproduced for many generations obviously store a priori knowledge which is valid.

The term 'a posteriori' is less useful. It seems to refer to a process of information gathering by observation and induction. No information transfer occurs in this fashion and there is, therefore, nothing the term refers to. However, the fine tuning which occurs in many organisms after birth is an a posteriori method of gathering knowledge, even though the method is not observational, but selective. The upshot of the discussion must be, therefore, that we distinguish not so much between a priori and a posteriori as between selectionism and instructionism. The method of information transfer is in all cases a selective method. There is an abundance of proposals on one side; and a selective process carried out by the other side. The world selects and the organism proposes. The proposals can be either in terms of genes, or in terms of synapses and nerve endings. In all cases, there is selection, not instruction. This is the distinction which matters and which we have to hold fast to.

We must think of evolution of living matter as a long-drawn-out process of knowledge acquisition by the natural selection of those organisms which are a fit to their environment. Such fitness has also been called adaptation. But, as we shall see, even 'fitness' is a very relative term; and 'adaptation' is misleading because it suggests that there is an absolute fit, the way a hand fits into a glove. This is, as I shall argue presently, not the case. It is more correct to think that the gene pool of every species contains a lot of information about a special part of the world which we call the environment of the members of the species. Evolution is the progressive selection of those organisms which, by their chance mutations, contain the best available information. Chance and necessity here come together in a single process. Mutations and proposals are random; but the selective process is anything but random. In principle, if one had sufficient independent knowledge of the world, one could predict the outcome of all selective pressures. The principle, however, is pointless, because there is no way in which one can predict what proposals will be made; that is, one cannot possibly know what these pressures will have to select from.

V

Each organism is a theory about its environment. This has to be understood, first of all, as a definition of the concept of environment. A frog which darts out its tongue at every bird that passes seems to have information about its environment, because the bird is certainly within walking or jumping distance of the frog's environment. But in a strict sense, it is not. The frog defines its environment as the subject matter of the theory or information it embodies. Since it is programmed to dart its tongue at small black moving objects (which we call flies), but not at large gliding moving objects (which we call birds), the birds are not part of the frog's environment. Biologists call the specific environment thus defined the organism's *Umwelt.* The concept of *Umwelt* was first introduced by Uexküll[35] half a century ago and has been used ever since in biology. But its philosophical significance and consequences have never been brought out. We shall return to this concept presently.

The statement that every organism is an embodied theory about its environment must be taken literally. The organism bodily represents information about its environment so that, at a pinch, one could learn a lot about the environment by looking at the organism. Thus, one can tell by looking at the gills of fish that they are a theory about an environment which contains oxygen, but not in the form in which oxygen is present in the air. We can also think of the organism as an embodied theory in the sense that it has not yet been falsified by its environment. In a Popperian sense, therefore, an organism is a provisionally true hypothesis. However, the environment is limited and specific, and the organism is a theory only about that specific part of the world which we call its *Umwelt.*

The term 'theory' in this context is to be taken literally. A theory states that one has certain expectations. An organism lives by having expectations. The frog lives by the expectation that flies will dart past; a bird lives by the expectation that its bones are light enough to enable it to fly, and a fish, by the expectation that the presence of oxygen in the water is such that its gills can pick it up. If and when these expectations are not met, the organism becomes extinct and the theory it embodies, falsified. For this reason we say that the knowledge stored by the organism is knowledge of regularities; not detailed knowledge about particular occurrences. An organism is not adapted to the myriads of events in the environment; but to the regularities which obtain in its environment. To quote G. M. Edelman:

> To be of adaptive value categorization must entail generalization, or the ability on the basis of a few stimuli to respond or recognize a much larger range of stimuli. To the degree that such generalization takes place, it allows the individual to deal with novel instances and to ignore other stimuli within a behavioral context ... it relieves the organism of the burden of storing large numbers of single instances.[36]

Furthermore, a theory represents parts or aspects of the world it is about. The shape of a fish and its neuronal reflexes represent water. They do so not by making a picture, a portrait or a double of water. Nor do they do so by describing water. It is indeed quite difficult to say what the term 'represent' precisely indicates. And yet, the behaviour of the fish and the functioning of a theory about water are exactly identical. The fish represents water by its structure and its functioning. Both features define an initial condition (e.g. the degree of viscosity of water) which, when spotted or sensed, triggers off a prognosis or behavioural response which, in the case of a fish, fails to be falsified. By contrast, a bird does not represent water. The initial condition it defines (e.g. a certain air pressure) will trigger a response which, if the bird is in water, will be falsified. Given these conditions of organisms, we can say confidently that organisms are theories. Since the theories are not expressed in words but in anatomical structure and programmed reflexes, we say that the theory is embodied.

The organism as an embodied theory represents that part of the world it is a theory about. But this representation is neither verbal nor conscious. Organisms act and produce goal-directed behaviour by self-organising processes long before they have evolved nervous systems. The nervous system is merely a more elaborate way of producing that behaviour. The paramecium and its phobic and topic responses, for example, is a perfectly good theory about water with a particular concentration of H-ions.[37] Similarly, the sheep tick is a theory about an environment which has a temperature of 37 degrees Celsius and contains buteric acid.[38]

When we speak of a theory, we are thinking of a general proposition from which one can deduce, with the help of an initial condition, a prognosis. An organism is an embodied theory precisely in this sense. It embodies general information about its environment, i.e. information about the regularities which obtain in the environment. When its senses spot an initial condition which is covered by the general information it embodies, the initial condition will trigger a response which is the precise parallel to a prognosis. Most philosophers have mistaken the observation which acts as an initial condition which triggers the prediction with the help of an existing theory for an observation, accumulations of which instruct the organism how to react and which thus form the foundation of the theory. Konrad Lorenz has provided ample evidence that these observations are triggers which release knowledge, not foundations on which knowledge is built.[39] When the frog sees a fly, this sensation is the initial condition from which it 'deduces' the prognosis that the tongue ought to dart out in a certain direction. The deduction is non-linguistic and a compulsive reflex. Human organisms are more complex. In one sense, they are embodied theories and make their deductions (reflex responses) like frogs and birds. In another sense,they are capable of language and formulate a special set of theories which are not embodied but expressed linguistically. In the latter case, the initial condition too has to be phrased as a linguistic expression. But this is a very special case which has its own problems as well as advantages, to which we shall turn presently. The pre-verbal organism can spot the initial condition as defined by the general information it stores without the help of words and produce the adequate prognosis without language. For an embodied theory (i.e. an organism) the presence of the initial condition triggers a certain response. The initial condition is not observed in the sense that the frog watches and then scratches its head to remember what it has done on previous occasions when a similar observation was made, trying to recall whether its behaviour on that earlier, similar occasion was successful or not. However, a frog whose tongue-darting behaviour is triggered off when the initial condition is an elephant embodies a false theory about its environment and such a theory will be falsified in the sense that such a frog will leave very few, if any, offspring.

The concept of 'initial condition' and the way it functions in relation to the theory is striking. The term is well known in both philosophy and science; but is always bedevilled by the problem of reference. How can one identify an event as the initial condition for a theory? How can one know what it refers to? Ideally, one should be able to do so independently of the theory and use the occurrence of the initial condition as a test for the truth or falsity of the theory. But such independent identification has proved very difficult, as numerous largely unsatisfactory theories of reference and meaning amply demonstrate.[40]

If, on the other hand, one starts with the theory and acknowledges that theories are not a posteriori – that is, put together by summarising or collecting observations of facts or objects – one will see that the initial conditions define themselves in terms of the theory they are part of. Take the frog as a theory. That theory will state that given the initial condition of a small, black object moving past at great speed, the tongue will dart out. There is no need here to look for the small, black object moving past at great speed and then check whether what is so observed will be the object 'referred' to by the initial condition. Having the theory in place, we can simply use the darting out of the tongue as the touchstone. If the tongue darts out, the initial condition is met: i.e. the object it refers to is indeed the object it refers to. Such a simple solution of the problem of reference is not possible when one believes that theories are a posteriori. But any a priori theory will automatically establish reference and solve the problem of reference implicitly. There is no need to know or define the object the statement of the theory means or refers to independently, so that one can compare the statement and the object and check whether it does or does not fit. The statement, on the contrary, predicts that an object will be found, and this prediction constructs or constitutes the object. If the thing that is predicted cannot be found, the statement refers to nothing. Since the statement, so to speak, creates the object it refers to, there is no need to look for the object's independent existence. Such reference by prediction from a priori theories is less inscrutable than the kind of reference envisaged by the more conventional theories of reference.

The parallel between embodied theories and linguistically stated theories will help. An embodied theory defines what counts as an initial condition. Frogs and birds have no problem of reference. Their perception of light on a surface is not a passive reception and does not stand in need of a theory of what it 'refers' to. Perception takes place *only* when it is expected. And what is expected (= initial condition) is defined by the a priori theory of the organism's programme. The world the frog's eye perceives is entirely and totally defined by the theory the frog's eye embodies. For the frog's eye, there is no world over and above its eye from which it might have to choose some features it ought to respond to. Its world is its *Umwelt.* I shall return presently to the circular relationship between theory and *Umwelt.* The *Umwelt* consists of nothing but initial conditions defined by the theory the organism embodies.

The difference between conventional philosophical understanding and the biology-oriented understanding of prediction and its relation to an 'initial condition' is well brought out when one examines the opening sentence of Quine's most recent book. 'From impacts on our sensory surfaces, we ... have projected a systematic theory of the external world. Our system is proving successful in predicting subsequent sensory input.'[41] Quine's entire premise is mistaken. Our system is not proving successful in predicting *any* subsequent sensory input. It is not, as he seems to imagine, standing ready, waiting for more sensory input. Our system, on the contrary, will venture a prediction provided the initial condition it defines is stimulated. It is incapable of making any prediction about any other conceivable sensory input. Quine goes on to wonder how we manage to predict subsequent input successfully. The answer is very simple: our system cannot help predicting those subsequent inputs which are identical with the initial conditions the system defines; and since those initial conditions are defined by the system, they must be successful predictions. Try to get a frog to dart its tongue out at anything other than a fast-moving object, the size of a fly!

This situation creates the illusion that the theory the organism embodies – or any theory for that matter – does not react to the world at all; but to a part of itself. H. R. Maturana[42] was therefore able to depict theories and organisms as if they were information tight in Ashby's[43] sense. He sees them as cybernetically governed by such signals as are contained in them, without reference to the outside world. A submarine, for example, Maturana says, is steered not by observations the captain makes of the ocean, but by meter readings of instruments inside the submarine. Maturana is perfectly correct in so far as the information that does come in is information defined by the submarine or the organism or the theory. But his picture is an illusion nevertheless, for it leaves one all-important fact out of account. The organism is a theory about the world which has become a fit to that world because of the organism's ancestors and because of natural selection. While it is true that at the present moment the organism appears to behave cybernetically because it defines its own initial conditions and is incapable of registering anything other than its own definitions as initial conditions, it is also true that such behaviour is correct or fitting behaviour because the organism has been selected naturally. Maturana's cybernetic explanation is viable because, and only because, the organism is the result of natural selection. Maturana's submarine is, in other words, doing what Kuhn calls 'normal science' – if the prehistory which produced submarine or paradigm is taken for granted, the pursuit of 'normal science' is self-validating and need be neither questioned nor explained.

Some years ago there appeared a cartoon in *Punch* which showed a kingfisher as an embodied theory, but it did so by way of a telling misunderstanding. The kingfisher was shown to be sitting on a branch above water, and in the water there was a fish eyed by the bird. Now, the kingfisher, when it plunges down to catch the bird, is able to discount the refraction of the light rays in the water and, in diving, makes allowance for the correct angle of refraction. In other words, the kingfisher correctly embodies a complicated theory about its environment. But the cartoonist, uninterested in our central contentions, added something which he ought not to have added. Above the kingfisher's head there appeared a bubble in which there was written 'Snell's Law'. Now, Snell's Law is precisely what an organism which is not an embodied theory about this environment would require. But the kingfisher, being an embodied theory, obviously does not need Snell's Law to direct its dive correctly. On the contrary, the verbal or mathematical formulation of Snell's Law is, if we are to continue to think in terms of the central contentions, a disembodied organism. Of this more later.

Each embodied theory or organism is adapted – whatever the precise meaning of adaptation is – to a specific *Umwelt.* The theory it embodies is a theory about that *Umwelt* and is not applicable to any other part of the world. As Uexküll explained, every organism filters out those parts of the world it expects. The eye of the frog identifies as initial conditions only changes in light and moving objects the contours of which are curved. Everything else which other theories 'know' about the world is simply neglected and relegated to oblivion.[44] This means that there are at least as many embodied theories around as there are species of organisms and that each embodied theory has its own *Umwelt.* The *Umwelt* is virtually and literally the creation of the organism. So while we say that the organism is adapted to an environment, this adaptation creates an abstraction of the environment which is the only part of the environment the organism knows. An organism which is not a frog can learn a lot about a certain environment by watching the frog – that is, by taking the frog as a theory about a certain environment. But the frog itself only knows that part of the environment in question which consists of the initial conditions which the theory embodied by the frog defines. This situation has a fascinating philosophical consequence.

From the frog's point of view, the frog is a theory about its *Umwelt*; and that theory is analytical in Quine's sense. 'Any statement', Quine wrote,[45] 'can be held true come what may [= analytical] if we make drastic enough adjustments elsewhere in the system.' In our case, the way the frog filters out a few features of the environment in order to create its *Umwelt* is precisely such a drastic change in the system which consists of the frog and the environment. As far as its *Umwelt* is concerned, the frog is a theory which is analytically true. Indeed, one must suppose that Uexküll's *Umwelt* is the answer to Bishop Berkeley's prayer: 'Please, Lord, let there be a world which exist, come into being and vanish, only because of perception.' Had Bishop Berkeley known more biology, he would have spoken of perception of an initial condition rather than of straight perception. But there is no denying that the discovery of the *Umwelt* is precisely the sort of world the Bishop had been looking for. When the organism no longer perceives its *Umwelt*, that *Umwelt* disappears without a trace. And, what is more, most organisms which thus create their own *Umwelt* are incapable of leaving a written record of what their *Umwelt* was like. Only in so far as they remain as fossils can we today form a conception of what their *Umwelt* might have been. However, from the point of view of a fish or of a *homo sapiens*, the frog is a theory about its environment, not about its own *Umwelt* which is its own creation and which does not exist for other organisms. To say that the frog is adapted to its *Umwelt* is a tautology; and when it is claimed that the frog is a theory, it is claimed that it is a theory about its environment, not about its *Umwelt*. The frog's *Umwelt* is merely the evidence the frog produces for the theory it is, and it is not surprising that the *Umwelt* validates the presence of the frog. What is surprising and substantial is the fact that the survival of the frog is, up to a point, validated or, as we should say more correctly, not falsified, by the environment it is living in.

There is again a telling parallel to conventional, linguistically expressed theories. According to Kuhn the facts, i.e. the *explananda*, are specific to each theory because their meaning is defined by the theory and does not exist independently. Hence it is, according to Kuhn, in principle and in fact, impossible to communicate across theories. The *Umwelt* of every organism is thus a perfect Kuhnian world which consists only of such facts or events as every organism specifies. The theory – i.e. the organism – which defines those facts also explains them and, in so far as the frog survives by darting out its tongue at small objects that fly past, it is doing 'normal science'. As far as organisms are concerned, Uexküll anticipated Kuhn by nearly half a century.

The philosophical interest of the concept of the *Umwelt* is urgent because there is indeed a parallel between embodied theories (organisms) and ordinary, linguistically stated theories. Every theory, and most certainly those mega-theories which Kuhn describes as paradigms, defines its own *Umwelt* and cannot be tested, let alone be falsified by 'facts' which are defined by a different paradigm. As far as organisms are concerned, Kuhn's theory of paradigms is a restatement of Uexküll's theory and as incontrovertible as Uexküll's original theory. *The Structure of Scientific Revolutions* should have been subtitled 'The Paradigm and its *Umwelt*'. But the matter cannot rest there. As far as ordinary, linguistically stated theories are concerned, we know that every single *Umwelt*, whether it is the *Umwelt* defined by an organism or whether it is the *Umwelt* defined by a proper scientific theory, must be compatible with every other *Umwelt*. There is only one universe, and it is inconceivable that different laws and different regularities should obtain in its several parts. It would therefore be a mistake to conclude that each *Umwelt*, while relative to a specific organism, is incompatible with or inconsistent with all the other *Umwelten*. On the contrary, since all possible *Umwelten* are part of the world, all different *Umwelten* must be compatible with each other. This does not imply that every embodied theory is compatible with every other embodied theory. A swallow is a theory about air; a fish is a theory about water. But if a swallow is thrown into water, it will, as an embodied theory, very quickly be falsified by the fish's *Umwelt*. So the swallow is, as a theory, not compatible with the fish as a theory, because the swallow-theory applies to a different *Umwelt* from the fish-theory. But the swallow *Umwelt* is compatible with the fish *Umwelt*, because both niches are part of the same universe. As we shall see presently, both the fact that each embodied theory is strictly applicable to one *Umwelt* and one only and the fact that all conceivable *Umwelten* must be compatible with one another even if the embodied theories about them are not, are of the greatest importance when we come to the evolution of theories which are not embodied as organisms.

For this reason, Kuhnian worlds, though quite real in the sense that they exist exclusively in virtue of theories, whether embodied ones or non-embodied ones, cannot be the last word. Once stated or evolved, they must be shown to be parts of a consistent system and must be shown to be compatible with one another. This may not always be possible. But in principle, the search for ultimate compatibility and the quest for the removal of contradictions is the goal, and this goal helps us to distinguish between theories. Theories which explain more and remove more inconsistencies are to be preferred to theories which explain less and leave more inconsistencies standing.

It is only meaningful to speak of a theory, embodied or not, if there is at least one part of the world to which it applies and of which it is true or by which it can be falsified. When we are thinking of organisms as embodied theories, we must be able to think of them as true – at least as true about one part of the world. We are wont to think of organisms as a fit to their environment or as adapted to their environment. If we are thinking of them now as embodied theories about that environment, we should be able to equate 'truth' with 'fitness' or with 'adaptation'. But we have to be very cautious here and sound a warning even though we are not prepared to abandon the concept of the organism as an embodied theory. In a sense we should be prepared to do so, once we recognise that adaptation and truth cannot be equated in a simple way. There is a very good reason why an adapted organism cannot simply be equated with a true embodied theory about a specific part of the world. An adapted organism is an organism which survives in a given environment. If there is a great deal of competition, the organism can survive only if it is very minutely adapted to the environment. In such a case it will fit the environment and we could think of it as an embodied theory which is true. But suppose an organism in an environment in which it has, for some reason or other, few or no competitors. In such a case comparatively little adaptation is required for the organism to survive, and it will be correspondingly difficult to think of it as a true embodied theory. In other words, the degree of fitness required for survival depends on the absence or presence of competitors. Under these circumstances a great many organisms fail to be falsified for cognitively irrelevant reasons. Hence no organism can be seen to be a completely accurate representation of the niche of the environment it is surviving in.

Organisms come into the world with almost all the information they need for survival ready present in the way they function. However, it is possible both to condition their responses and to imprint them. These ways of learning and picking up information after birth must not be mistaken for an ability to pick up additional information from the environment. A dog can be conditioned to have its saliva flow at the sound of a bell, provided it has been exposed for some time to the sound of the bell every time it has been given food. In this way, it has been shown that a dog can be fooled to behave in a stupid way, because the dog cannot observe correctly that there is a causal link between saliva and food, but no causal link between saliva and the bell. If the sounding of the bell was arranged by a benign Skinnerian and will always be accompanied by food, no great harm is done. The Skinnerian, contrary to his intention, has simply demonstrated that dogs cannot learn from observing the environment. If they could, their saliva would soon stop flowing when the bell was made to sound without food. But suppose the food and the bell happened by coincidence over a long period of time because the dog was living near a church, and suppose the kind owner one day disappeared and the dog was left without food but with the bell. The saliva would still flow precisely because the dog could not observe what was going on and could not put one and one together.

Konrad Lorenz's discovery of imprinting comes closer to our conception of taking in instructions from the environment than Skinner's conditioned responses. Lorenz found that newly hatched mallard ducklings will accept as their 'mother' the first object that quacks and waddles in front of them. Lorenz concluded correctly that the ducklings carried quite specific information about what is their mother from the moment of being hatched, that is, without ever having experienced a mother. They are, however – and this is consistent with the theory of neuronal fine tuning – a little uncertain as to which precise object of the several objects they will encounter is their mother. They are born with the expectation that their mother will quack and waddle in front of them. Any quacking and waddling object will be the initial condition they expect, and that initial condition will trigger their behaviour of following. Once their neurons are finely tuned to that initial condition, it will be virtually impossible to alter their responses.

VI

So far these contentions have been fairly uncontentious. I have merely introduced some new terms like 'embodied theory' for well-known phenomena and processes. But when we now turn to *homo* more or less *sapiens*, the contention that theory formation continues the process of evolution will become more daring and more novel and will be, at first sight, less acceptable. Indeed, when we come to *homo sapiens*, philosophical interest in the growth of knowledge is shifted from the evolution of *homo sapiens* as an organism to *homo sapiens* as the organism which produces linguistically expressed theories. The evolution of knowledge continues; but where at first it developed in the shape of embodied theories, it now continues in the shape of disembodied organisms. There is a perfectly good reason, as we shall see presently, why we should consider linguistically expressed theories as disembodied organisms. To put it briefly and to anticipate, linguistically expressed theories are subject to the same process of evolution by chance mutation and selective retention – even if the selection will turn out not to be entirely natural – as organisms or embodied theories. For this reason it is useful to refer to them as disembodied organisms. This is not just a matter of analogy; and we are not simply saying that theories are *like* organisms. For we have seen that one could just as readily consider organisms as theories. As the analogy is symmetrical, we can confidently assert that theories *are* disembodied organisms and that organisms *are* embodied theories.

The organism of *homo sapiens*, like all other organisms, is a theory about the world. But as an embodied theory it has now for millions of years shielded itself more and more from natural selection by the environment and does not, as an organism, represent the world in which it is surviving in a very truthful way. Even pre-human organisms, I have argued, often have a low degree of adaptation to their environment because, accidentally, there may be a comparative absence of competitors. By contrast, the human organism has made a fine art of shielding itself against competitors by using a specific strategy for doing so, and there is nothing accidental in the construction of that shield. Hominids – even though we cannot pin-point the moment at which this process began – prevented an abundance of proposals to the environment from which the environment might have selected the most fit, by sexual repression. At the same time, as offspring became purposefully less abundant, hominids further subtracted themselves from purely natural selection by the ingenious device of institutionalising cooperation (only rudimentarily present among non-human primates) through social solidarity which, in contrast to the minimal solidarity of herds and troupes of pre-hominid animals, lasts over many generations. Such solidarity has to be organised on the basis of non-natural criteria, such as the performance of rituals or the maintenance of beliefs. Thus any selection which takes place inside a given society proceeds by promoting organisms which perform well in these rituals and which espouse whatever belief is current, rather than by promoting individuals whose behaviour is adapted to the natural environment. Such institutionalised solidarity, moreover, compensates for any deficiencies in organic structure and in truthfulness of belief systems. For it makes it possible for any group who, for example, are unable to grow enough food because they are deficient in bodily health or deficient in their understanding of climate, to form a war-gang who will steal food from other groups better adapted to their environment. In this way, the human organism has not only successfully subtracted itself from purely natural evolution, but has also, by evolving traits which are adapted to its social rather than to its natural environment, more than sufficiently compensated itself for lagging behind in biological evolution. Since we are concerned here with the evolution of knowledge, the human organism would cease to be of interest, were it not for the fact that it is possessed of consciousness. While it loses interest for philosophy as an embodied theory, it becomes a new focus of interest because its consciousness enables it to produce disembodied organisms.

The term 'disembodied organism', though curious, is appropriate, because, as will be argued, consciously proposed and held theories show remarkable likeness to the embodied theories we have been talking about and, moreover, though powered by a different mechanism, develop and evolve according to the same principles. If consciously held theories are referred to as disembodied organisms, they will be seen, correctly, as objectively existing entities. This amounts to a reification and accords with Popper's view that consciously held theories are like objects. Hence, Popper[46] speaks of 'epistemology without a knowing subject'. With such reification, we are also close to Hegel's concept of 'objective spirit'.[47]

At first sight, one might expect that the emergence of consciousness makes it possible for the deficiencies of biological evolution to be corrected. We noted that pre-human animals have a very restricted possibility of learning from experience and that the truth of the theory they embody can often, even though they survive, be very limited or that the theory they embody can be a less than accurate representation of their environment. When consciousness is present, one could expect that these deficiencies be remedied. Consciousness, one could surmise, will make learning by taking instructions possible; and will make it possible for the organism to observe the environment accurately in order to produce theories in which the way the world presents itself and the way it is represented by the theory are made to coincide. It has been held that the ability of consciousness to observe accurately so that learning by following instructions becomes possible gives organisms with consciousness an evolutionary advantage and that that advantage is the reason why the phenomenon of consciousness has been selected for survival. But this is not so. This interpretation of the mysterious phenomenon of consciousness is a misinterpretation.

For example, when biology started to be taken into account by philosophers, there was a brief moment when it came to be believed not only that the natural selection of all sense organs provided a guarantee that organisms would be helped by their senses to survive, but that one could also build a science of science or a theory of knowledge which stretched beyond what was immediately present to the sense organs, on the adaptiveness of sense-organs. This looked indeed like a positivism revitalised by biology, a positivism with a new, evolution-based, lease of life – a proof of Locke by Darwin! 'It seems to me possible', Abner Shimony wrote in 1971,[48] 'to present considerations that reduce the arbitrariness of choice among frameworks' and that make our preference for one system of cognitive coordinates more than a purely pragamatic preference. 'The evolutionary point of view', he continued on the next page, 'supports the causal theory of perception by providing quasi-teleological explanations for many features of the perceptual powers of human beings and other animals, thereby answering one of the challenges of Berkeley.'

Shimony, it seems, was proceeding too fast. He is right as far as organisms before consciousness and three-dimensional language are concerned. Here, the natural selection of sense-organs has produced embodied theories about the environment. *Homo sapiens,* in so far as he or she is an organism, is a similar embodied theory. If one followed Shimony's argument, it would lead one to the following conclusion. In human animals, natural selection has produced an organism the sense-perceptions of which are not totally veridical because all natural selection usually leads to survival in spite of less than perfect adaptation. However, in human animals, consciousness, having been naturally selected and being more plastic and responsive than mere sensation, must be the faculty which can correct everything in mere sensation that is less than purely veridical. But such a conclusion is incorrect. The presence of consciousness and of its direct consequence, three-dimensional language, greatly reduces the human animal's usefulness as an embodied theory while, at the very same time, enabling it to sprout disembodied organisms in the form of linguistically formed proposals instead. These latter proposals, like proposed organisms themselves, emerge before exposure to the world and are not causally induced in the organism by the world, no matter how adaptive man's five senses are. This is just as well, because, as we have seen in Chapter 1, there is a hiatus between a linguistically formulated theory and any causal impact of the world. The non-conceptual world cannot be conceptualised without further ado, and such conceptualisation as emerges is either over- or underdetermined by the impact of the world on our senses. Contrary to Shimony, Darwin does not rehabilitate Bacon and Locke and is no comfort to positivism; for consciousness cannot be used as a simple, adaptive device to straighten out any deficiencies in fitness likely to result from natural selection.

The reason why consciousness is nevertheless an adaptive advantage and has been selected is more complicated and very devious. We have seen in Chapters 1 and 2 that consciousness is not a faculty which makes accurate observation possible. On the contrary. The presence of consciousness introduces an element of uncertainty and unease into the behaviour of the organism. To start with, if we saw that there is a causal link between the impact of the world on the organism and the emergence of consciousness, we also saw that the feeling which thus emerges, and of which we are aware, is undefined and inchoate. It cannot therefore enable us to make any representation until it is articulated. The articulation, however, is hypothetical and can never be more than hypothetical, because the feeling in question cannot be referred to unequivocally over and above being so articulated. This means that it is not possible to state whether any articulation of the feeling is true or false, for there is nothing unequivocal enough to compare the articulation with and to which it can be seen either to correspond or not to correspond. Consciousness, therefore, cannot be used as a corrective for the deficiencies of natural evolution. Its uses are indirect and more oblique, and it is because of the results of such obliqueness, rather than because of its allegedly direct advantages, that consciousness has an adaptive advantage.

The first and immediate result of the inarticulate consciousness of feelings is, as was argued above, an unease, a lack of orientation. Consciousness, in its subjective immediacy, is an almost uncomfortable feeling of unfocused awareness. This unease encourages the transformation of two-dimensional methods of giving messages about what is the case into three-dimensional expressions about hopes, fears, the future, the past, possibilities, and so forth. The most important characteristic of human, three-dimensional language is, as was argued, the ability to express states of affairs which have not been observed and which do not exist or occur. Thus it is possible, using three-dimensional language, not only to lie and to express errors but also to make statements which are unwarranted by evidence or experience. In the class of these expressions we find both false and true statements, but for which there is no compelling or cogent reason regardless of whether they are true or false. Both are possible, and exclusively possible, in three-dimensional language.

The importance of three-dimensional language for the evolution of disembodied organisms (theories) cannot be exaggerated. It would indeed be theoretically possible for theories to be formulated without three-dimensional language. Such theories would simply report what the observed – whatever precisely this may mean – information instructs one to report and would not go beyond the information given. In the middle decades of our century, stemming from the efforts of Carnap as well as of Ryle and Austin, attempts have been made to prune all language down to the expression of information that was given or reduce language to statements that do not go beyond the information given and hold all language that cannot be so reduced as 'meaningless'. If these attempts had succeeded or at least proved more convincing than they have, it would have been impossible to account for the development of knowledge expressed in sentences by the selective process of evolution. For evolution proceeds by selection; and unless there is an abundance of theories to select from, evolution cannot take place. But 'abundance' here means presence of theories which are false and theories which are true, so that selection can take place. But without three-dimensional language it is impossible to formulate statements which go beyond the information given and which can, therefore, be false. In any case, both true and false statements have one thing in common: both go beyond the information given. One cannot distinguish the true from the false by saying that the true ones are the theories which cling to the information given and the false ones, the theories which go beyond the information given. It is just as well that three-dimensional language *is* available; for the phenomenon of consciousness and the potential for articulation makes it – as we have seen – impossible to stick with the information given and behave like a pre-human animal in which responses are reflexes. The human organism, that is, cannot confine itself, except in special cases, to the unworded impact the world makes on the organism.[49]

The primary importance of consciousness consists in the fact that it promotes three-dimensional language; and the importance of three-dimensional language consists in the fact that it makes possible the proposal of both true and false theories. These theories compete with each other and are subject to selective pressures. As a result of the operation of these selective pressures, knowledge grows.

In order to explain why three-dimensional language produces both false and true theories indiscriminately and why consciousness (the sponsor of three-dimensional language) does not play the guiding role philosophers have been fond of imagining,[50] but an oblique and devious role in the growth of knowledge, we have to look more closely at what false and true theories have in common. The two statements 'this sheep is green' and 'all sheep are mammals' have one feature in common. Both go beyond the information given. The first statement has a colour predicate which goes beyond the information given. The information given would require the predicate 'white'. In the second statement it is the word 'all' which goes beyond the information given. It so happens that the first statement is false and the second, true. Or consider the General Theory of Relativity and the theory that all birds have three legs. They have something in common. They go beyond the information given. The Theory of Relativity goes beyond the information given and certainly did so when it was first proposed, and the second theory goes beyond the information given in that it contradicts immediately observational data. Nevertheless, no matter how much they have in common, there is a clear difference between the two theories. But this is a finding which comes after the proposals and which is the result of critical selection. For such selection to take place, we first have to have, if not an abundance, at least two statements, one of which can be not selected. But in order to get an indiscriminate set of statements or proposals, one has to have a language which allows the proliferation of false statements. Only a three-dimensional language permits such proliferation. Without proliferation there would be no evolution of knowledge. The indiscriminate abundance of theories made possible by our ability to produce statements which go beyond the information given repeats the wild proliferation of organisms on the level of biological evolution. If the first organisms had produced nothing but clones, there would have been no evolution and living matter would still consist of nothing but an abundance of identical protozoa or whatever it was which first took shape in the primeval soup.

However, at this point, we need a pragmatic consideration. When evolution switches from organisms (embodied theories) to theories (disembodied organisms), the theories continue to be a priori, in the sense that they come *before* the organism which produces them has been exposed to the world. In principle, proposals of disembodied organisms could be random. All that is needed for competition and for the selective mechanism to operate is that there should be at least two different such disembodied organisms. But in practice these proposals of disembodied organisms are not random. They tend to come in clusters so that we can speak at any one time and in any one place of a 'culture' which exhibits a fair degree of homogeneity. Such clustering is due to the fact that in any one place at any one time, there is a finite number of problems (and observations) which will promote a set of either similar or directly contrary, rather than a proliferation of random, proposals. Moreover, the emergence of such proposals, like organic evolution itself, is a form of *bricolage*.[51] Any new emergence, be it of embodied theory or disembodied organism, takes place through a reshuffle or recombination, no matter how faulty, of some of the structures which are already available. The DNA of an elephant is no more likely to make a mistake of duplication which results in the production of a mouse than a body of religious beliefs about a law-like universe is likely to inspire a sudden proposal that the universe is a chaotic occurrence. None of this has any bearing on the process of selection, which will operate regardless of whether the proposals come in clusters or are random. All we need to notice here is that the proposals are most unlikely to be wild-cat. I shall return to this topic in Chapter 5.

Two-dimensional languages can be used to state some falsehoods and thus go beyond the information given. But the range of such falsehoods is very restricted. In a two-dimensional language one can only say that something which actually *is* the case is *not* the case; that is, one can deceive or mislead. But one cannot make positive statements about fabricated events, one cannot invent things that are not the case and, in short, make statements about an alternative to reality and thus go beyond the information given.[52]

Natural selection is only possible because of mistakes and depends on mistakes. If DNA always reproduced itself with unfailing precision, there would have been nothing to select from because there would not have been different organisms. The advantages of sexual reproduction, whatever its pleasures, consist in the fact that it provides a greater variety of offspring for the environment to choose from than reproduction by simple cell division. We have seen that the human organism shields itself through sexual repression and communal solidarity, no matter what mistakes there are in reproduction, against natural selection by the environment and thus perpetuates a large number of traits which are non-adaptive. As mere animals, human beings are not well adapted, and if a human organism were suddenly to be left to fend for itself outside a community (even if one male and one female were so left outside), offspring capable of survival and reproduction would not stand much of a chance. When we are looking now for the continuation of natural selection in human organisms, we must not look towards the biological organism itself, but to the theories it expresses in language; for it is in these theories, rather than in the organism itself, that the mistakes from among which selection takes place occur. Since these theories continue the process of knowledge acquisition by chance mutation and retentive selection, they are to be seen as disembodied organisms. Embodied theories (= organisms) and disembodied organisms (= theories proper) have one thing in common. They go beyond the information given and thus provide an opportunity for selective pressures to operate and differentiate among them.

Popper[53] has drawn attention to one immediate advantage which disembodied organisms – though he does not use this expression – have over embodied theories. For natural selection to operate on embodied false theories, one has to wait for a whole generation to die out. But the selection of disembodied organisms can take place very much faster, because theories expressed in language can be dropped long before the organisms which hold them die out. Evolution can therefore proceed at a faster rate when the organisms to be selected are disembodied. The main advantage, however, is that consciousness promotes the kind of three-dimensional language in which it is possible to formulate theories about states of affairs which do not occur as well as about states of affairs which do occur, but for which there is no immediate evidence. Thus, for example, a three-dimensional language can express a belief that the soul is immortal and that the earth is flat as well as the belief that the sun will always rise, even though the only evidence available tells us that the sun has always risen in the past. In other words, consciousness is for making mistakes *via* the language it promotes. It is not for making the way the world presents itself coincide with the way we represent it, i.e. it is not for getting things right by following instructions. It helps us to get things wrong and to produce an overabundance of proposals, regardless of whether they are right or wrong, so that the better-fitting ones can be selected for retention and the rest be made to fall by the wayside. Consciousness is necessary, not because, as philosophers ever since Plato have held, it acts as a methodical guide to truths and goodness, but because it is able to generate wild guesses.

VII

Here we encounter an immediate difficulty. If we accept – as we do – that information is not transferred from known to knower by instruction (so that the known instructs the knower and the knower picks the information up by listening or looking as carefully as he or she can), we must also accept that information transfer takes place by selection from an abundance of proposals. Selection is a straight alternative to instruction. As far as the natural selection of organisms or embodied theories is concerned, we can suppose more or less accurately how it operates. Surviving organisms represent a comparatively truthful description of at least their immediate environment, and their survival vouchsafed that kind of truthfulness. The selection for such survival was natural and automatic. In biological evolution truth simply emerged relentlessly and inexorably. The sorting out of conjectures (embodied theories) is done by differential birth and death rates. Obviously, the same cannot be said for disembodied organisms (= theories). To be sure: when there are competing theories, we can select some and reject others. But there can be nothing automatic and natural in this process. On the contrary: the absence of natural selection suggests that in this case, the process of selection is arbitrary, and this indicates that any truthfulness which emerges as the result is also arbitrary. Better, such truthfulness is not really worthy of the name.

The difficulty appears quite formidable. To start with, disembodied organisms have to be espoused by somebody. These somebodies must live in societies, and the survival of these disembodied organisms depends in the first instance on the fate of the organisms which hold them. Their survival, in other words, is governed by extraneous and cognitively irrelevant factors. The process by which they frequently survive is indeed so devious that in innumerable cases, the cognitively most useless and untrustworthy theories survive best. This will become clear after a little sociological reflection.

One of the many devices available for social bonding is a shared belief. Such a belief can act like a charter and guarantee the continuity of a social order and the solidarity and cooperation of its members over long periods of time. But the employment of beliefs (or, in our terminology, of theories) for such non-cognitive purposes produces a downright irrational method of selection of beliefs. A true belief can and will be taken up by every person of sound reason and good will. It can therefore not be used as a basis of a social bond; for one of the important functions of a social bond has to be the ability to exclude persons from a given solidarity system. The bond has to be such that it sets boundaries. The definition of outsiders and their exclusion by such a definition is in fact one of the most important ingredients in the cement which makes up the bond. When sociobiologists made the important discovery that human beings, to be capable of forming societies, must be genetically programmed for a certain amount of altruism, they forgot to mention that by the same token, they must also be genetically programmed to recognise limits to such altruism because altruism, like patriotism, is not enough. Social human beings have to know where their altruism has to stop. By the nature of the case, there can only be one set of true beliefs; but there is available an infinite number of false beliefs. A false belief which, because of its special form of falsity, is specific to a particular group, is obviously more capable of providing this kind of social bond than a true belief. A false belief, unlike a true one, can furnish an efficient criterion of exclusion. For this reason, false beliefs have a great survival value.

Their survival value is further enhanced by the effect they produce. On the face of it, one would suppose that a false belief will soon secure its own falsification, because people who espouse it will behave so stupidly that they will not survive for long. If the belief in question is, for example, that grain grows when one prays and not when one tends one's fields, one will suppose that the people who harbour such a belief will soon starve to death. But this supposition is mistaken. The false belief promotes cooperation and solidarity among the people who are using it as a catechism. Such cooperation will make them into a formidable military power and enable them to take food from the people they conquer. The people with the false belief will, therefore, not only not starve to death, but will suppose that it is their false belief which makes them thrive; for it enables them not only to get all the necessary food, but also to enjoy the pride they derive from their ability to conquer. The false belief, therefore, has a very good chance of survival because one of its effects is to shield the people who harbour it from the inexorable consequences which would normally accrue to stupid people who cling to false beliefs.[54]

Purely biological evolution is a relentless process of knowledge acquisition, even though, no matter how long it continues, the knowledge is piecemeal and specific to the *Umwelt* of each embodied theory and not an accurate representation of that *Umwelt*, depending on the accidental rate of competition the embodied theory happens to be exposed to. If competition could be absolutely perfect, the resulting representations would presumably be more accurate. But once we come to *homo sapiens*, the relentlessness of the process is stopped. First, *homo sapiens*, as an organism, has managed to evade purely natural selection by developing social systems in which the weak are protected artificially and in which qualities which are cognitively not directly adaptive are selected. In some societies, the selection of cognitive faculties is exceedingly devious, as when organisms are selected for survival because of their ability to fashion disembodied organisms. But even this oblique approach to knowledge is not enough to secure the continuation of the process of knowledge acquisition, because pieces of false knowledge can get kidnapped and selected for survival because of their solidarity-promoting quality. We find, therefore, that a very special social order is required for disembodied organisms to remain exposed to competition and the selective process which is necessary to assure the emergence of truthfulness among them. Free competition among theories is possible only in social conditions in which solidarity and cooperation can be assured without using theories for non-cognitive purposes. Such societies are generically referred to as 'open societies'. They hold together because people have nothing better to do or because their members are bonded to one another by a recognition of mutual self-interest. They are, for these reasons, non-catechismal and, as far as knowledge is concerned, cognitively neutral. In honour of Adam Smith, who was the first philosopher to understand their functioning correctly, we could also call them Smithian societies.

Truly Smithian societies are few and far between and, for this reason, the process of knowledge acquisition is as intermittent and discontinuous as one would expect. And, as if this were not enough, we now come up against the last and most fundamental difficulty. Even in a Smithian society, where theories are in free competition, there remains a grave problem. What, precisely, is the mechanism of selection? Since selection is not automatic, it must be done by somebody. When it is *done*, rather than happens, it must be done according to criteria of criticism. In an influential article on this matter, R. C. Lewontin[55] has raised what he thinks is the fatal question: 'What is the *basis* of judgement?' The question is well put and crucial, because, if the selection is not automatic, there must be a criterion of selection. Unless it can be answered satisfactorily, the contention that the evolution of theoretical knowledge is as Darwinian as organic evolution must be abandoned.

Lewontin points out that Popper tends to avoid a precise answer to this question, even though it is clear that he postulates an argumentative function by which critical selection and selective retention are carried out. Lorenz, according to Lewontin, has faced this question more squarely by arguing that natural selection is not only the moulder of embodied theories, but also the moulder of the human mind which puts forward disembodied organisms or theories. If one could imagine that the generation of language-based theories about the world (i.e. of disembodied organisms) were itself a sort of mindless interplay between the world and what Lorenz calls 'the mind' which has evolved, nothing further need now be said. One could then simply take it for granted that the human mind, being the product of natural selection, would not generate theories or views of the world which are grossly inaccurate or non-adaptive. Natural selection has left the human mind with certain abilities and dispositions of reasoning and thought capable of sizing up the environment correctly. Organisms, including *homo sapiens*, with minds and reasoning powers unable to do so, could not be here today, for their ancestors would not have survived. For Lorenz, there is no need for selection of theories because false theories could not be produced in the first place and competition could not arise. But this is not so. The world is silent and theories about the world are anything but silent. There is a hiatus somewhere, and in the bridging of this hiatus, as we have seen, there is endless opportunity for error of transcription. It is indeed the prevalence of error which provides competition between theories and which makes it necessary for us to answer Lewontin's question. Lorenz's answer is not an answer, but an attempt to explain that the question cannot arise because the human mind, as the surviving product of evolution, is well-nigh infallible; or, at least, as infallible as the judgement of the baboon as it swings from branch to branch or of the kingfisher as it dives for fish. Not only is this not the case, but – as we have observed before – man's ability to make mistakes has not only one, but two different and contrary adaptive advantages. First, it provides a proliferation of true and false theories which are in competition and subject to selective pressures: the ability to make mistakes explains why there has been a *growth* of knowledge. Second, false theories, better than true theories, function as social bonds and provide the basis for the solidarity and permanently structured social cooperation without which the human animal, depleted as it has become of so many natural instincts, could not survive. Where Popper was too vague, Lorenz is evasive. Lewontin's question remains unanswered.

Lewontin's question can only be answered by two further contentions. These contentions are not arbitrary, but derived by a non-Kantian transcendental deduction. Kant's famous transcendental deduction yielded the twelve categories in terms of which all knowledge had to be stated. But the deduction was a deduction from, so to speak, Newtonian premisses. Whatever its merits or demerits, it lost its cogency when it turned out that Newton's was not really the last word about the universe. The central thought of Kant, however, seems beyond reproach. When we are looking at a house which is standing up, it is legitimate to deduce that there must be foundations, even though the foundations themselves are not visible. Following Kant, we can make a deduction from the fact that we have knowledge. But, unlike Kant's, the deduction will not be made from the assumption that the universe is Newtonian; but from the fact that evolution has taken place. If it had not, we would not be here to wonder about it. The fact that evolution has taken place entitles us to deduce two characteristics which the universe must have. First, there must be regularities. Organisms which are adapted must be adapted to regularities. Second, all regularities must be compatible with each other. Any one embodied theory or any one disembodied organism may be adapted to a specific part of the universe, and that adaptation may not be compatible with another adaptation. A fish is adapted to water, but not to air. In a sense, a bird is incompatible with a fish, unless one is able to be neither bird nor fish and have a disembodied organism (= theory) which explains how water and air can coexist in the same universe. Moreover, a sheep tick, for example, is able to respond to the temperature of 37 degrees Celsius and to buteric acid. Its presence and survival are predicated on its assurance that the temperature and the acid must be compatible. The compatibility of regularities must, therefore, be a feature of the universe. Third, it follows from the second contention, though not from the fact that evolution has taken place, that disembodied organisms (theories) will tend towards progressive elimination of incompatibilities, so that the evolution of knowledge will appear to be unilinear rather than cumulatively circular. Such unilinearity continues the unilinearity of cladistic taxonomy for biological organisms in which species always diverge and never converge again.[56] Though birds and mammals are descended from a common ancestor, they will never again be able to breed a hybrid. In the same way, it is inconceivable that we will ever take up Ptolemy's universe again. This corollary to the second contention has an important implication for the history of science, because it provides a criterion as to which of competing theories we must prefer.

Larry Laudan has put it very well: 'My proposal will be that rationality consists in making the most progressive theory choices, not that progress consists in accepting successively the most rational theories.'[57] In other words: progress happens because it is rational that we should always prefer that theory which explains most. It does not come about if we decide to give preference to the most 'rational' theory because there is no way of telling which of several theories is the most rational one.

The non-Kantian transcendental deduction yields a minimal ontology which helps us to find the answer to Lewontin's question as to the mechanism of selection because it provides a directive for belief revision. It sets up some kind of Archimedean point from which to judge beliefs and theories because it is, itself, outside the belief system.[58] When we are confronted by competing theories, we single out those theories for selective retention which best live up to the minimal ontology. The requirements of the minimal ontology cannot be arbitrary, because they are derived from the fact of evolution itself. They are a guide as to what we must expect of a true theory, because we cannot retain as true any theory which does not conform to the minimal ontology. The process of selection, though not automatic, is not arbitrary. It must be controlled by the requirements of the minimal ontology. And since these requirements correspond to the basic characteristics of the world, they are not arbitrary, and the non-natural, critical selection which the competing theories undergo is, therefore, quite natural after all.

What precisely do these requirements amount to? They amount, first, to the notion that only a theory which makes a statement about regularities and invariance can be a candidate for selection. A theory about a unique concatenation of two or more events is to be admitted provisionally only until a theory in which some of these events appear as regular occurrences is found. We critically scan the field of competing theories and eliminate all those which have nothing to say about regularities. Once theories about regularities are retained, we then continue the process of selection by applying the criterion of falsification to them. This will lead to further selections and narrow the field.

Second, there is the requirement that all selected theories must be consistent with one another. This guides our selection towards the elimination of theories which are inconsistent with one another. For example, the caloric and the kinetic theory of heat are not compatible. After critical examination, we select the kinetic theory, because it explains not only the behaviour of gases, but also the behaviour of a great many other things. The preference for the kinetic theory removed contradictions. The selection was not guided by the social fact that the kinetic theory has more adherents or by the cognitive consideration that it is more elegant; but by the consideration that it is more likely to correspond to a world of which we know that it cannot contain inconsistencies. Thus our selection is guided by the notion that the fewer theories we are left with, the closer we are to a proper representation of the world. Ideally, one single theory should explain everything, in the sense that lots of lesser theories which explain particular occurrences ought to be deducible from it. The very notion that knowledge should *explain* is derived from the idea that we should try to remove contradictions. It also follows from this requirement that in all cases we must prefer those theories which explain more. Thus, for example, we prefer the General Theory of Relativity to Newton's theory of gravitation, because it explains not only all the occurrences explained by Newton's theory, but also a great many other occurrences. For this reason we say that it is a better explanation and prefer it and, by doing so, eliminate its competitor.

The answer to Lewontin's problem is crucially important. It not only establishes that selection continues even though it does not do so naturally and automatically, but also establishes that when selection takes place according to the requirements of the minimal ontology, we can solve a problem which stands at the centre of both Kuhn's and Quine's philosophy of science. In these views, paradigms and even major theories are changed for all sorts of reasons. But confrontation with nature is not one of them. 'There can', Kuhn writes,[59] 'be no comparison with nature.' Quine puts it slightly differently.[60] He argues that the bottom line for science is a fit of patterns of sensory stimulation to theory. But these sensory patterns are always described theoretically – they are theory-laden. Hence, the bottom line is the comparison of theory to theory. At no point can one break out of the prison-house of theoretical terms. Science can only be understood from within and there can be no vantage point outside theory from which we can test whether a theory fits any data. These views conform to the view that initial conditions are constituted either by embodied theory or disembodied organism and that they, together with the theory, only produce nothing but those predictions which are made by the theory. As in Kuhn and in Quine, here too, one cannot break out of the prison-house of theoretical terms and confront facts or nature or whatever. If this is really the last word, knowledge will not be worth pursuing. But our answer to Lewontin shows that this is *not* the last word.

The answer to Lewontin shows, on the contrary, that the criteria which dictate our preference for theories, though not derived from our observation of facts, are not arbitrarily chosen. It is, therefore, not true that we cannot break out of the prison-house of theoretical concepts. While it is true that we can never confront a theory with anything raw and non-theoretical, we can compare theory with theory and must give preference to those theories which explain most. This preference criterion is not arbitrarily prescribed,because it is formulated in harmony with what we know of ontology. By using the criteria we obtained from the non-Kantian transcendental deduction, we are providing a substitute for confrontation with raw facts. While any one theory, at best, can only fit something that is theoretically described, we must always prefer that theory which explains more than another. In this way, knowledge ceases to be a pure convention and we escape from the Kuhn-Quine Ferris Wheel. We do so not by contradicting them and by maintaining that theories can somehow be confronted with raw nature, but by showing that there is a factor which they have overlooked: that is, they have overlooked the non-Kantian transcendental deduction. However, since such selection of theories is artificial rather than natural and automatic, one could use other, non-cognitive criteria of selection. Artificial selection could indeed give, and is known to have given, preference to theories according to aesthetic or political criteria. If such criteria are used, the selected theories cannot have cognitive value.[61] Only the criteria derived from the non-Kantian transcendental deduction can give cognitive value to artificially selected theories.

If our knowledge of a minimal ontology is the basis of the mechanism of selection, we must take another look at the problem of induction. The existence of regularities is taken for granted, even though we may be ignorant as to which regularities actually hold. Induction has become a problem because it was recognised that no number of particular observations can establish that any one regularity actually is holding. The most it can prove is that the observed instances are the case, but this does not make a regularity. Philosophers have always attributed enormous importance to induction because without it, they have believed, the advance of knowledge would appear a miracle and, for this reason, they have spent an inordinate amount of time in attempts to show that, against all reason, induction is a rational method of discovering regularities. No amount of virtuosity has ever made any of these logical acrobatics in the least bit persuasive. In any case, the problem, as Karl Popper showed as far back as 1934,[62] disappears as soon as one recognises that the rational process of discovery is not an inductive process. However, the recognition that the fact that evolution has taken place establishes that there must be regularities in the world, could be taken to mean that hopes are raised again. For Jonathan Cohen argued in his *The Dialogue of Reason*[63] that induction is justified as long as we can have prior knowledge of the existence of uniformities. With this argument Cohen moved the question from the logical acrobatics of writers like Stove[64] to ontology. It had been claimed for a long time that the practicality and validity of induction depended on the assumption that there *are* uniformities in the world. But so far such an ontological assumption hung in the air. Now, in taking evolution into account, such an ontological assumption ceases to be a mere assumption and becomes an ontological fact. Does this mean that induction is justified after all?

The answer must be a plain negative. If one is an instructionist and believes that knowledge is picked up by piecemeal observation, stored and then generalised, one is not entitled to believe in evolution by natural selection and, therefore, the transcendental deduction which led to the conclusion that there are regularities because evolution could not have taken place without them, cannot be carried out. Instructionists can have no knowledge that there are regularities. On the other hand, people who accept that evolution has taken place do not hold that knowledge is acquired by instruction; and to them, the newly raised hope that inductions are practicable and valid after all is of no use. According to them, knowledge is not acquired by induction, and the hope that it might be is redundant. But there is more. It is one thing to assert that there really are regularities and to infer from this assertion that theories about reality ought to state regularities and that theories which state the most regularities should always be given preference. In this procedure, the knowledge that there are regularities is used as a criterion of preference; not as a basis for induction. It is quite another thing to assert that there are regularities and that, therefore, it is logically valid to conclude from a limited number of regular occurrences that this particular series of regularities can be projected into the future. After all, any one particular series of limited occurrences may be a false series. The upshot is that, though evolution justifies the belief that there are regularities, this belief can only be used to establish a criterion of theory preference. It cannot be used as a basis for the induction of any particular regularity. In the traditional view, induction was considered a rational way of finding general theories and the practice of induction was considered to constitute reason's way of finding the truth about the world. Clearly in the new, biology-oriented view, there is no such role for reason. But this does not mean that reason is eliminated from the knowledge-acquisition process. It is merely transformed from a positive faculty which finds the truth into a negative, critical faculty which eliminates errors and, above all, those errors which consist in the presence of false theories and of theories which, though not false, explain less than other theories. Rational intellectual behaviour is now seen as error elimination rather than as truth-finding ability. In this negative and critical role, it is an essential part of the growth of knowledge. The only rational behaviour which is now recognised is disrespectful, unlimited and uninhibited criticism. By contrast, the old role of reason is now dismissed as having played an irrational role.

VIII

In order to answer Lewontin's question we had to search for a non-natural and non-automatic mechanism of selection which was not arbitrary, but based on some features of the universe. It will have been noted that we have advisedly omitted truth as a possible basis. Truth is a vague concept which becomes thornier and thornier the more one tries to make it less vague. In a general sense we think that a statement or theory is true if it corresponds to the facts. But the word 'corresponds' begs the question. Does correspondence mean that the facts are correctly described in words? If so, what is the meaning of 'correctly' when the description is in words and the state of affairs alleged to be 'described' is not in words? Tarski's famous definition of truth by correspondence is not helpful, because it is confined to the relationship between two types of sentences. On one side he has the statement 'It snows today' and, on the other, the statement that it is snowing. If they correspond, he is saying, the first sentence is worthy of the adjective 'true'. I do not think that anybody would quarrel with him; but it is clear that this definition is confined to a definition of the relationship between two types of sentences. There is nothing in it which helps us to break out from the bonds of language and confront language with reality. One cannot belittle Tarski's achievement. As Hilary Putnam pointed out, it was Tarski who managed 'to separate the problem of truth from the problems of the nature of knowledge and the nature of warranted belief ... [and showed that one could have] a theory of truth which is neutral with respect to epistemological questions and even with respect to the great metaphysical issues of realism versus idealism'.[65] But both Tarski's limitations and his achievement merely highlight the age-old problem of truth: it is comparatively easy to arrive at a concept of truth. But as soon as one seeks to define the precise criteria by which one can determine whether any particular statement is true or not, one is confronted with an array of possibilities, none of which are satisfactory. This impasse can be no surprise. For 'truth' is supposed to be a relationship between a linguistic expression and a non-linguistic event, that is, between two conditions which are not really related.

For this reason it is recommended that truth be deleted from the list of criteria of selection. What we are really looking for are criteria by which we can compare linguistic statements. The linguistic statements imply statements about initial conditions, and such implications determine and define what the initial conditions refer to. In our theory that theory formation is a priori, an independent problem of reference cannot arise, and the tantalising question what it is that expressions refer to, over and above the initial condition that is designated by the theory, is obviated. The question whether the initial condition that is designated by the theory refers 'correctly' to an independently ascertained state of affairs cannot arise. For this reason, we compare theories with other theories and concentrate on criteria of comparison rather than on the question as to what it is that a consequence of the theory 'refers to'. With this view we are not siding with a simple coherence theory of truth according to which it is relevant whether theories cohere, rather than whether they refer. On the contrary. We continue with a theory of correspondence, but have managed to let the problem of 'correct reference' take care of itself. The criteria necessary for comparison of theories are provided by the principles we have from our non-Kantian transcendental deduction as to what reality is like. Thus we know that there must be regularities; and that there must be consistency. The former principle guides us to select those theories which assert regularities, and the second principle guides us to select those theories which remove contradictions.

Once the process of selection according to these non-automatic but non-arbitrary criteria is in place, it must proceed in a unilinear way. This unilinearity of the growth of knowledge amounts to progress, in the sense that it moves away from a condition in which we have lots and lots of particular statements, many of which seem incompatible with one another, to a situation in which we get more and more statements about regularities, most of which are compatible with one another. This movement is not to be thought of as an attempt to move towards a predetermined goal, but as a movement away from a primitive condition. The further we move in this direction, the greater the fit of our knowledge to the world we are living in. Knowledge as an embodied theory is applicable only to one specific *Umwelt.* But knowledge expressed as disembodied organism, once the process of selection is under way, is selected and retained as knowledge because it explains, among other things, how all the different *Umwelten* are compatible with and related to one another.

All this shows that no concept of truth need be set up as a primary concept. But one can form an idea of truth *via* the explanatory power of theories. Theories which explain lots of things are more likely to be true than theories which explain only a few things. The greater the explanatory power of a theory, the greater the likelihood that it is true. It is, of course, logically conceivable that a theory with very great explanatory power is not true or less true than a theory with little explanatory power. But for a theory to explain lots of different things and yet not be truer than a theory which explains only one of these things, one would have to suppose that a colossal accident has happened. One would have to assume that there are massive coincidences, if a theory which explains lots of things is not true or *less* true than a theory which explains only one thing.

In the mind of Larry Laudan[66] the persistent attempt to take truth as a primary concept has finally reduced itself *ad absurdum.* He writes that 'for every successful theory in the past of science which we now believe to be a correct theory, one could find half a dozen once successful theories which we now regard as substantially non-referring'. Laudan's observation is perfectly correct. But his conclusion that therefore theories can never be taken to refer to reality or be regarded as true does not follow. It would only follow if one takes 'true' as a primary concept. But if one takes 'truth' to be derivative from the growth of the explanatory power of theories, the conclusion no longer follows. For then one can recognise that theories once held to be referring and now regarded as non-referring, have suffered their fate not because of 'having been true' and now 'no longer being true' – which would indeed be paradoxical. They have suffered their fate because they had low explanatory power to start with and have been superseded by theories which have greater explanatory power. The application of the term 'true' is almost accidental. Many of the earlier theories 'referred' correctly and are even today referring correctly. However, they referred to very few things, while the theories which have superseded them refer to a great many more things.

The notion that knowledge acquisition progresses by this method of selection from a large number of proposals allows us to see this progress as yielding more and more verisimilitude. The concept of verisimilitude was first introduced by Popper in order to distinguish better theories from merely good theories. Some theories, he argued, were better than others because they were a better approximation to the truth. Hence, he argued, we should distinguish theories not so much according to their truth and falsity but according to their verisimilitude.67 The criterion of verisimilitude was introduced because it was recognised that the whole and entire truth could never be reached and that the theories which are often preferred do not necessarily, in one fell swoop, 'correspond' to reality. Popper explicitly stated that this is the way we look at theories intuitively and how we compare them. 'To say that the aim of science is verisimilitude has a considerable advantage over the perhaps simpler formulation that the aim of science is truth.'68 However, understandably, he wanted to justify and formalise this intuition. Unfortunately, Popper decided to derive his formal statement of verisimilitude from a Tarskian definition of truth, and there he ran into considerable difficulties.

Popper correctly suggested six criteria for distinguishing good theories from better theories.[69] Two of the six concern matters of precision; and they go without saying. The other four criteria state that theories are preferred according to their explanatory power. Again, nobody would question these criteria. There is, it must be noted, no reference to truth in any of the six!

But then, on the following page, Popper jumps ahead and defines verisimilitude in terms of truth and content. He starts by saying that a true statement can have, as its logical consequences, only true statements; but that false statements may (his actual word is 'will', but I take this to have been a Freudian slip!) have both true and false conclusions. And then, at the end of this paragraph, he continues: 'Thus whether a statement is true or false, *there may be more truth, or less truth, in what it says,* according to whether its content consists of a greater or lesser number of true statements.' Having established by some jump that true and false statements are symmetrical in that they *both* can have both false and true consequences, he arrives at his definition of verisimilitude: 'the *verisimilitude of a statement will be explained as* increasing with its truth content and *decreasing with its falsity content*'.[70] If a statement has more false than true consequences, that is, it is less verisimilitudinous than a statement which has more true than false consequences. As one might expect, there have been countless criticisms of this doctrine of verisimilitude.[71]

However, in the heat of these criticisms it has always been overlooked that Popper's fault was in his reasoning, not in his doctrine. He ought to have arrived at the doctrine and at his recommendation that we see verisimilitude rather than truth as the aim of science by a different route. What is more, had he remained with the six criteria of comparison,[72] he would have arrived at the doctrine without the faulty reasoning he committed himself to because of his inordinate respect for Tarski's definition of truth. It was this doctrine which tempted him to formulate the important concept of verisimilitude in terms of the respective sizes of the truth content and the falsity content of theories.

The reasoning should have gone something like this. A true theory cannot have false consequences. Therefore one cannot compare true theories according to the sizes of their falsity content. At most, one could compare false theories according to the sizes of their truth contents, and in this case one would have to say that some theories are more false than others. But this is far from Popper's definition of verisimilitude. However – and this is all important – two true theories can be incompatible with each other. Every theory creates its own *Umwelt*; and every such *Umwelt* must be compatible with the environment it is part of, even though and especially because the theory says nothing explicitly about the environment, but only defines an *Umwelt.* The theory, however, does not represent the environment or describe it correctly for the following reason. For any one theory, there are many others which, though equally compatible with the same environment, are not compatible with each other.

To explain this astonishing situation, let us make a thought experiment. Suppose we have a frog and sheep tick which could tell us what the initial conditions are of the theory they embody. Suppose we make them sit, one after the other, in exactly the same place so that we can be sure that any difference in their *Umwelt* is not due to a difference in their perspectives. When questioned,the frog and the tick will make incompatible claims about the same spot in the same environment. If we suppose further that in front of them there is a sheep and a buzzing fly, the frog will then claim that there is a buzzing fly and a lot of empty space; and the tick will claim that there is a sheep and a lot of empty space. Though these make incompatible claims, the frog-theory will verify its claim and the tick-theory, its claim. Or, better, the buzzing fly and the sheep with its buteric acid and a temperature of 37 degrees Celsius will fail to falsify the frog-theory and the tick-theory. There is a very simple explanation of this state of affairs. Each theory, by implying certain initial conditions, stakes out precisely what it refers to. We do not observe the fly and the sheep independently of the frog-theory and the tick-theory and then wonder how the frog and the tick, respectively, manage to get their initial conditions to 'refer' correctly. The buzzing fly and the sheep are being referred to correctly because they are the initial conditions the two theories imply. If there were no buzzing flies and no sheep, the two theories would be false, i.e. the frog and the tick would not have evolved and would not be here.

In order to find out what the environment is really like – as distinct from the two incompatible representations – we need to work towards a theory which will make the two *Umwelten* coincide. It is doubtful and indeed improbable that a theory which will produce a complete coincidence can ever be found. But we may conclude that the greater the coincidence between any theory's *Umwelt* and the environment, the more likely it will be that the theory will correspond to reality. Hence a theory in which the coincidence of *Umwelt* and environment is greater than in another theory will have greater *verisimilitude* than the other theory.

Seeing that we can never know independently what the environment is like, how can we measure the varying degrees of coincidence? There is a very simple answer, and it is contained in the four of Popper's six criteria which deal with degrees of explanatory power. The answer is that there is greater coincidence in *that* theory which explains *both* the frog-theory's *and* the tick-theory's *Umwelten.* Such a theory will remove the incompatibility between the frog and the tick. Thus the greater the explanatory power of a theory, the greater the likelihood that it will correspond with reality and tell us what is really going on. And this, we conclude, is what is meant by 'verisimilitude'. One can apply the same reasoning, *mutatis mutandis,* to Newton and Einstein. The conclusion will be identical with the conclusion we reached when we compared the frog and the tick. We prefer Einstein's theory to Newton's theory because the former, in explaining both everything that Newton's explained and a great many other things as well, removes more contradictions than Newton's theory.

The evolution of living matter as well as of theories stands in a cognitive relationship to the world. We could think of evolved living matter as a sort of gnososphere which surrounds the world. This knowledge would not have evolved had there not been a real world which progressively and incessantly selects organisms and the basic features of which are used to select theories. Knowledge comes either as embodied theories or as disembodied organisms. For the former, the process of selection is automatic and natural; for the latter, the process of selection is not automatic but, once in place, is carried out according to non-arbitrary criteria which reflect the fundamental qualities of the world in which the process is taking place. The knowledge which evolves in this manner is knowledge of a real world, and for this reason, when biology is taken into account, we must decide in favour of a philosophy of realism. However, this realism is conditional and hypothetical, for the knowledge in question is knowledge due to selection. We know that that knowledge would be different if the world were different; because if the world were different, it would have selected different embodied theories, and we would have had to select different disembodied organisms. But the selection process is not perfect and allows the survival of theories and organisms which are not a perfect fit. Hence we call this realism hypothetical realism in order to distinguish it from common or vulgar realism.

In conclusion I have to draw attention to an important and telling difference between Kant and Popper. Following his reasoning, Kant had to come to the conclusion that the knowledge of regularities we have is 'prescribed' to the world. From his point of view this conclusion and the choice of the word was inevitable. Our forms of perception and the categories of our understanding are, in Kant's view, a priori to our experience of the world; but are not related to the world in any significant way. Hence the knowledge which we base on them is 'prescribed' to the world. What is more, the world can never be experienced in a way which could conceivably make us or it rebel against these prescriptions. For all possible experience is filtered through the same channels through which we arrive at the prescriptions. If the world rebels, we will never find out! The Kantian world must be a tyrant's dearest wish-dream. Popper and Lorenz have helped us to transform this Kantian transcendental idealism into hypothetical realism. For Lorenz takes into account the fact that, since we have evolved, the forms of our perception and the categories of any understanding are directly applicable to the world because they themselves have evolved by natural selection. Popper, further, takes into account that the process of knowledge acquisition continues beyond the evolution of organisms by non-arbitrary critical selection. Therefore, whatever is known because it has survived the selection processes is applicable to the real world, but only hypothetically so, because the information about regularities that is stored has been stored by selection from a wide variety of proposals and could, probably as well as conceivably, be replaced. The selection assures that it is applicable to the real world, but not that it is the sort of precisely accurate representation of the real world which we might have obtained, had we been able to learn from instructions issued by the world. In Popper's terms, the knowledge we have is therefore not prescribed to the world, but proposed to the world; and, more precisely, it consists of those theories which are left standing after the process of selection by criticism is temporarily exhausted. In this way biology has transformed transcendental idealism into hypothetical realism and replaced Kant's *prescriptions* by Popper's *proposals*.

The change is by no means just a matter of semantics. It leads us right into the heart of the problem of realism. The belief that the world 'really' exists and any view of what it might 'really' be like has always been taken to depend on one's ability to establish what it is like when nobody is looking; what it is like in itself or, as Hume put it in Appendix I of the *Inquiry Concerning the Principles of Morals*, how things 'really stand in nature'. On the face of it, the problem does not even make sense, for 'looking' is obviously an animal, if not exclusively human, category and the idea of what anything might 'look' like when nobody is looking is patently absurd. So absurd, that no matter how one turns the problem around, one is always left with a Kantian agnosticism in regard to noumena, to the world as it-is-in-itself.

The hypothesis of evolution changes all this. There is, to begin with, a striking contrast between the science of physics and the science of biology. By no stretch of the imagination can one suppose, nor has anybody ever done so, that physics tells us what the world is really like. Even when physics was simply the study of matter in motion, physicists approached that study from at least two different and incompatible ontologies – the Cartesian world, in which everything was full of matter; and the Newtonian world, in which there was space, with matter floating around inside it. Whichever way, for a long time materialism was considered the hallmark of reality, and a realistic understanding of the world required a reduction of everything that is happening to material events. But the more 'matter' was studied, the more elusive it became. Physics, in spite of its successes, cannot tell us 'how it is'. Its findings are appearances, mere meter-readings, instrumentally or operationally interpreted, verified by other meter-readings. The situation changes radically when we turn to biology. It is not that biology necessarily provides correct information about how life functions. On the contrary. As far as its specific findings are concerned, it is in exactly the same boat as physics, and its advances consist in many ways in the application of physics and chemistry. But in studying living matter and its relation to the non-organic world, that is, with the theory of evolution, biology is leading us very directly to a viable ontology by providing a picture of living matter as a genuine representation of what the world is really like. Take, for example, the category of 'looking'. 'Looking' refers to the ability to make use of photons. If there were no photons, the ability to 'look' would not have evolved. The fact that it has evolved indicates that there are photons and that with the help of these photons, there is something to be seen. The category of 'looking', in other words, would not be available if there were nothing to be looked at. This realisation still leaves us with the problem of hallucinations and optical illusions. But it does point to the fact that when we are asking what the world 'really' is looking like, we are not moving in a closed space of a mere appearance, to be distinguished from the true reality of the noumenal world. In a nutshell: the world of living matter which has evolved is not a mere appearance, but tends to present itself to itself as it really is in the shape of embodied theories and disembodied organisms. Living matter is the world as it appears to itself, not as it appears to an observer. With the hypothesis of evolution, the distinction between noumenon and phenomenon has become unnecessary.

Donald Campbell[73] has warned that this conclusion should not be mistaken for a claim that all living matter is so precisely adapted to its environment that one can learn from it what the noumenal environment actually is. On the contrary. There are many successful strategies which result in organisms which do not embody noumenally correct information. We need to make a subtle distinction here. The world of living nature is the way the world of non-living nature reveals itself to itself. Adaptations which look to us as if they were indirect or vicarious cannot be thought to be phenomenal or anything other than noumenal. However, it is true that we human beings, as onlookers, frequently fail to understand the noumenality of the world revealed by the behaviour and existence of living matter. And, therefore, it often looks to us human beings as if the adaptation is not a revelation of the noumenal world. But this comes from the fact that we seem to have some weird, privately independent conception of the noumenon – quite mysterious and the more mysterious for the fact that we can have no conception of it other than how it reveals itself in the world of living matter. And when we find that organisms do not quite live up to what we suppose noumenality should be, we believe, as Donald Campbell does, that noumenality is not revealed after all. My argument, however, is that it is manifest; and that it merely looks as if the old dichotomy of noumena and phenomena should be retained, because it is we who are failing in our understanding of the manner in which living nature represents the world.

5

THE VIEW FROM SOMEWHERE

I

From the perspective of biology, knowledge – that is, the cognitive relationship between two terms – is a form of self-reference. There is the world; and there is the world in a pattern shaped by the world itself, and that pattern refers cognitively to the world. The knower is part of the known and has been shaped by what is known. The reflector reflects, more or less adequately, because it is itself part of what is being reflected. The biological perspective, therefore, provides an assurance that the reflector is adequate and also explains, at the same time, how it has been shaped by natural selection, to be adequate. As Carl Sagan once remarked, 'humans are the stuff of the cosmos examining itself'.

There has always been enormous doubt in this area. One of the central problems of philosophy has been to determine how adequate the reflector or knower or mirror can be thought to be. Historians of philosophy are all agreed that the problem stems from Descartes's concentration of a knowing subject which faces an objective world. By thus formulating the knowing relationship as if the knower were some alien from outer space or a person supposedly occupying an Olympian stance, Descartes made it well-nigh impossible to account for the adequacy of the knower, the reflector or the mirror. Whether the problem originated with Descartes or not, without biology it remains a very formidable problem. For wherever one thinks the knower is taking up his position, he or she has no intimate relation to what is known, and whatever he or she is asserting cannot be considered to be adequate to the world as long as one cannot show that the knower or the mirror itself has its origin and place inside that world and is part of it and that the relationship of knower to known is a relation of self-reference.

There are good reasons why the question of adequacy should be in the centre. Whatever is paraded as knowledge must claim, among other things, that it is equipped to refer to the world. There must be indications as to how it is to be tested, what theories are to be given preference, what kind of coherence with other knowledge is to be expected and what theories are to be counted as corroboration. A Platonist will expect this adequacy to be defined in terms of the universality of the concepts used; and a Kantian will define this adequacy in terms of the formal categories of understanding such as causality, modality, and so forth. In order to satisfy oneself as to such adequacy, one has to know, more or less, what the world is like. But such a requirement leads to the paradoxical situation in which one is supposed to know already what the world is like before one can determine whether whatever one knows about the world is adequate to the world.

There is no need to examine again the simplistic argument advanced by positivists that the adequacy of knowledge is derived from the method of its acquisition. If we hold still, they have been saying from Locke to the Vienna School, the world will impress itself on us and form us in its own image. We have seen that, given the nature of our nervous system and its relation to language, this view is spurious. The opponents of positivism, from Plato to Kant, have, however, not come up with a satisfactory explanation of adequacy either. Let us go over their arguments again. Plato believed that the knowledge we have of the world is adequate because it is vouchsafed by the ideas or forms which some kindly world-soul or demiurge has made ready to be inspected by us before our birth. If there is an inadequacy, so much the worse for the world's ever-changing and unstable components. In the middle ages St Thomas Aquinas argued that there must be adequacy because our minds and the world have been made by one and the same Creator. Though deeply indebted to biblical mythology, strangely enough, St Thomas divined that knower and known are of the same stuff and that, therefore, the possibility of some kind of *adaequatio* can be taken for granted. In more modern times, Kant reasoned that adequacy is assured because evidence which might cast doubt on adequacy is perceptually inaccessible and can therefore never intrude into the world we know. In our own century, the search for adequacy seems to have been finally abandoned. Instead we are invited by Thomas Kuhn to be satisfied with some kind of pragmatism which takes knowledge to be a convention which temporarily recommends itself because it solves some problems while leaving others to be solved by some other convention at some other time. In this pragmatism, the question of the correspondence of a theory to the world does not arise.

From this vantage point, Richard Rorty has argued that Descartes and Kant were very misguided in thinking that if they could only polish the mirror more carefully, it would yet yield adequate knowledge. Neither Descartes nor Kant can be said to have polished the mirror. Contrary to Rorty, they merely strove to define its limitations. If they were misguided it was not because they kept polishing the mirror to make it reflect more accurately, but because they failed to see that the mirror owes its existence to the world and is made of the same stuff as the world and that the cognitive relationship which exists between the mirror and the world is like a cybernetic system in Gotthard Günther's sense; or in the sense of Max Delbrück, who shows that *res cogitans* and *res extensa* are made of the same stuff and do not oppose one another. We must not think of that cognitive relationship as a relation in which an Olympian subject confronts the world. It is, so biology has made us realise, a relation in which the world of living matter mirrors the world. Living matter has evolved in the world it mirrors and owes its ability to mirror – even though this ability is, as we have seen, not perfect – to the world it mirrors.

Before biology was taken into account, philosophical reasoning, as soon as it was deprived or had deprived itself of the fideistic scaffolding of religion and mythology which St Thomas Aquinas had used to generate confidence in the adequacy of our knowledge, invariably increased not only philosophical but also ontological anxiety. This was due to the fact that philosophers took physics to be the model of knowledge, and were looking exclusively at the universe of physical events. If one looks at the world picture of physics one will not find a subject or a subjective consciousness. Physics is how an atom appears to an atom; not how it appears to a conscious subject. The atom or electron does not store information about other electrons in its environment. Its relation to other electrons is not a cognitive relation, but a causal relation. This is not altered by the fact that in the course of the history of physics, causality has been understood in many different ways – as push, as pull-and-push, and as propensity.[1] Whatever electrons are doing to each other, they are not responding to each other as if they had information about each other. The relation of electron to electron is not a cognitive one.

In such a picture, if one thinks of a knower, that knower must decidedly be an outsider, somebody who is trying to watch from a cosmic vantage point. He or she is looking down like an alien from outer space and does not belong to the system he or she is trying to know. But since there is no outer space such an alien could come from, such a view is literally a view from nowhere,[2] which must leave the adequacy of his or her knowledge always in dispute. In this picture, *res cogitans* cannot be part of *res extensa,* and the adequacy of knowledge is defined in terms of the separation of *res cogitans* from *res extensa.* This is not to say that physics yields nothing but a world of matter behaving in a deterministic manner. On the contrary. We have come a long way from the physics of Laplace. Physics has introduced us to indeterminacy, to propensities and, above all, to 'matter' which is more like spirit than like the matter of common sense. Physics has also introduced us to entropy and irreversible unilinear development, so that cosmogony has become an integral part of physics, where in classical times there had only been laws of physics like Newton's laws and Maxwell's equations, which were time reversible. But none of these developments and none of the changes in our physical picture of the world has altered the requirement that the objectivity of knowledge depends on the fact that the knower should *not* be part of the world he or she knows.

The world described by physics is a world in which all relationships are described in terms of how an electron appears to another electron. There is explicit exclusion of an observer, other than other electrons. If and where an observer intrudes, the findings are labelled as subjective distortions. In this world there is no cognitive relationship, other than the relationship between the physicist on one side and the way the electron appears to an electron on the other side. But the relationship between the electrons is a non-cognitive, purely causal relationship. The physicist, who is one term of the cognitive relationship, stands nowhere. The objectivity of our knowledge of physics is constituted by the fact that the knower stands nowhere, and above all, not in a place which could affect his or her knowledge. While this position of 'nowhere' guarantees the objectivity of knowledge, it also shows up the arbitrary nature of the cognitive relationship. The physicist is an outsider who imposes himself or herself on the way an electron impinges on other electrons. There is no guarantee that the cognitive relation of the physicist to the electrons is adequate. As long as observers are outsiders, there must remain legitimate doubts as to whether what they are watching – in this case, how electrons impinge on other electrons – is really all there is to be 'seen' or all that is taking place. Unless the observers can be seen to be an integral part of the system observed, their view, no matter how objective, will be the view which appears to somebody standing in their place, even though this place is nowhere in particular and certainly just because it is nowhere inside the system. The only adequate view is the view from inside the system, that is, the view the system has of itself. Hence, the philosophical anxiety whether one has got it right or whether one ought to look from somewhere else continues to hold sway.

Attempts to use physics to prove the contrary have always failed. Ever since the Special Theory of Relativity and more so since the discovery of entropy and quantum mechanics, it has been surmised and argued that the observer's position is essential to the phenomena described. It was suspected at first, for example, that the Special Theory of Relativity was based on the fact that we have no way of measuring simultaneity, rather than on the fact that simultaneity does not take place, and even Einstein is known to have toyed with this possibility. Entropy too, it is widely believed, is not so much a fact of nature, as a result of our inability to have sufficient information about the movements of molecules, that is, it depends on the observer's ignorance. And in quantum mechanics the Kopenhagen Interpretation holds that the quantum jumps of the electron are determined, after all, by the observer's intrusion. But every one of these attempts at a subjectivisation of physics, which resulted from the view that an observer plays a crucial role in the physical world, has been more or less successfully resisted in the name of a preference for the view from nowhere. For only such a view is objective. The observer, in classical physics as well as in post-classical physics, must remain an outside observer, and his or her presence must not be thought to disturb the way electrons impinge on electrons. If, on the other hand, the cognitive relationship does not take place inside the world of physics, it is irremediably subjective. When we are left alone as outside spectators, we can have no confidence in the adequacy of our knowledge and must be left with the eerie suspicion that what we are watching is not the real world. This is the root of philosophical anxiety. 'We are like sailors', Otto Neurath wrote many years ago,[3] 'who on the open sea, must reconstruct their ship but are never able to start afresh from the bottom.'

This philosophical anxiety is readily transformed into genuine ontological anxiety. We wake up in the dead of night from a deep sleep and are aware, and for a split second we are nothing but pure awareness. Suddenly we gasp with horror and astonishment when we discover that we have a body which is impenetrable to our hands, which resists being pushed and pummelled, which is hard and occupies space. It is there. As we rally to full consciousness, we realise that we are suffering from Alzheimer's disease. We have no idea how we got here, how our ancestors got here, why we are here and why we are doing what we are doing. Any momentary explanation that comes to mind fills a gap, but in the end there remains a blank and a void. The sheer horror of realising that we are here but have no idea how we got here and why this hard impenetrable mass which we are and which surrounds us is the way it is becomes unbearable. Here is Bertrand Russell, writing about it in 1902:

> Man is the product of causes that had no prevision of the end they were achieving ... his origin, his growth, his hopes and fears, his loves and beliefs, are but the outcome of accidental collisions of atoms ... Only within the scaffolding of these truths ... can the soul's habitation henceforth be safely built ... A strange mystery it is that Nature, omnipotent but blind, in the revolutions of her secular hurryings through the abysses of space, has brought forth a child, gifted with sight ... with the capacity of judging all the works of his unthinking Mother.[4]

And here is Jean-Paul Sartre, half a century later, writing about the nausea induced by the contemplation of the blackness of a chestnut tree:

> [It is] like a bruise or a secretion, like an oozing – and something else, an odour for example; it melted into the odour of wet earth, warm, moist wood, into a black odour that spread like varnish over its sensitive wood, in a flavour of chewed, sweet fibre.[5]

Or look at Joseph Conrad in *An Outcast of the Islands*:

> A half-naked, betel-chewing pessimist stood upon the bank of the tropical river, on the edge of the still and immense forests; a man angry, powerless, empty-handed, with a cry of bitter discontent ready on his lips; a cry that, had it come out, would have rung through the virgin solitudes of the woods as true, as great, as profound, as any philosophical shriek that ever came from the depths of an easy-chair to disturb the impure wilderness of chimneys and roofs.

Such philosophical and ontological anxiety is not new. Hume suffered as much as Conrad, Sartre and Russell and gave full vent to his despair at the end of the first book of the *Treatise*:

> The *intense* view of these manifold contradictions and imperfections in human reason has so wrought upon me, and heated my brain, that I am ready to reject all belief in reasoning, and can look upon no opinion even as more probable or likely than another. Where am I, or what? From what causes do I derive my existence and to what conditions shall I return? … I am confounded with all these questions, and begin to fancy myself in the most deplorable condition imaginable, environed with the deepest darkness, and utterly deprived of the use of every member and faculty … If we believe that fire warms, or water refreshes, it is only because it costs us too much pain to think otherwise … Where reason is lively, and mixes itself with some propensity it ought to be assented to. Where it does not, it never can have any title to operate on us.

Hume accepted that the sentiments he had expressed sprung up naturally and stated that if he tried to banish them, 'I *feel* I should be the loser in point of pleasure.' Russell, proud and autonomous, gloried in the defiant freedom of his vision; Sartre, subdued by the realisation of the horror, succumbed to nausea. But these personal reactions are matters of taste, perhaps of stamina. The ontological anxiety is common to all these writers. But if one examines these passages critically, one will see that something jars. Hume had no idea why the warmth of fire and the refreshment provided by water should yield pleasure; and when he professed that pleasure is a better guide to philosophy than abstract reasoning, or rather, that abstract reasoning must be a function of pleasure – 'where reason is lively and mixes itself with some propensity, it ought to be assented to' – he did not suspect that if it were not for such and similar pleasures, we would not have evolved and would, therefore, not be here. He thought, on the contrary, that we are condemned to follow the lead of pleasure because we are ignorant of why we are here. Ignorant of evolution, he could not envisage the necessarily intimate link between pleasure and our existence. Or take Russell's statement that blind nature has brought forth a child gifted with sight. A reader with a knowledge of biology will immediately realise that if there are photons in nature, the presence of children with sight cannot be an accident. Beings with a sensitivity to photons have been naturally selected, because in a world in which there are photons, the ability to be sensitive to them is an adaptive advantage. Biology makes us realise that the contingency of human existence, which is the central point of these statements, is not entirely final and absolute.

There is no need to labour this point. One must grant that there are many reasons for ontological anxiety. The question, however, is how much ontological anxiety is reasonable and justified; and at which point it becomes neurotic, because unjustified. Since anxiety is unpleasant, it is worth asking how much of it is justified. We should reformulate and apply Occam's Razor: *anxietas non est multiplicanda praeter necessitatem.* What, then, are the limits of this necessity? How much anxiety is justified?

II

As soon as we add biology to physics, our grasp is radically altered. 'Physics', writes Gerald Edelman, 'is not enough.'[6] With the advent of living matter, biology has introduced us into a world which is full of self-reflection, because living matter, selected to be adaptive to its environment, reflects both living matter and non-living matter. Through chance mutation and selective retention, the physical world has generated a biological world in which one part of the world reflects another part. In this world there are relationships which are not purely causal – be it in the sense in which causes are pushes, pulls-and-pushes or propensities – but cognitive. In this world, one part contains knowledge about the world because living matter evolves by the acquisition of knowledge. If there are photons and light, there will be parts of the world which are sensitive to photons and light. Or, to use the terminology of the preceding chapter, there will be parts of the world in which there are theories about photons and light. Biology has opened a new perspective because, biologically, the eye is not caused by photons, but selected by photons. Our knowledge of light ceases to be a knowledge from nowhere, and Neurath's mariner, obliged to rebuild his ship in the middle of the ocean, will be able to avail himself of this cognitive relationship without imagining that he is pulling himself up by his own bootstraps. He can base himself instead on the firm knowledge that the link between photons and cells that are sensitive to photons is adequate and necessary and anything but a contingent, let alone conventional, relationship.

The knower is now seen to stand in a necessary relationship to the known. In biological perspective, the knower is no longer an observer from outer space, looking upon a world in which there is nothing but swirling electrons or sub-atomic particles. The knower, is, on the contrary, an integral part of the known system. For living matter to emerge and then evolve by chance mutation and selective retention of adaptive features, the rest of the world has to be in a certain condition and able to furnish the raw material for this emergence. There have to be a large number of chemical elements other than hydrogen, and the development of these elements depends on the amount of time available and on the nature of the forces operative in the universe. Nuclear forces, for example, are just strong enough to tie protons to neutrons. If they were a little weaker, no element other than hydrogen could be in the universe and the emergence of living matter would have been impossible. If the electromagnetic forces were only slightly smaller, all stars would radiate their energy as hot, blue objects in no time at all and there would have been no time available for the evolution of living matter. If, on the other hand, that force were a little greater, there would have been no stars with planets for life to have evolved on.[7]

A word of caution is needed when we are linking the world of dead matter with the world of living matter. In the past it has often been considered that the subject matter of biology must be *sui generis.* It must either have been created by God or be spurred by a vitalistic principle or some other kind of breath of life. On this assumption, the attempt to link dead and living matter was always considered an attempt to reduce living matter to nothing but the mechanico-causal motions of dead matter. But, as Ilya Prigogine observed, now that biology is beginning to understand how dead matter has transformed itself into living matter, the boot is on the other foot.[8] In linking biology and physics we are not demoting biology to the level of inert matter, but we are raising the world of dead, inert matter to a new, hitherto unrecognised, level, because we can now see that it must always have had the potential for giving rise to living matter. The present argument is to be understood as an argument to evaluate this potential, and not as an argument to reduce living matter to the level of dead matter. Instead of reducing biology to physics we might say, somewhat rhetorically, that we are now elevating physics to biology.

In order to find out the precise meaning of Prigogine's formula, we have to take a closer look at the relationship between physics and biology. The question of this relationship has been much debated and leads us to a matter of principle. Is biology an autonomous science, or is it mere province of physics and chemistry? At least this is the stark opposition which Alexander Rosenberg has discussed in his *The Structure of Biological Science.*[9] 'Autonomists' are people who believe that biology is and must be insulated from the distinctive methods of the physico-chemical sciences, and 'provincialists' are people who believe that biology is a part of physics and chemistry and that its theories must not only be compatible with the latter but 'actively' cohere with the latter. In this sense, provincialists are really reductionists. Though Rosenberg finds faults with both these views,[10] he concludes that those parts of biology which cannot be accommodated in principle to physical science should be jettisoned. In this sense, he decides the debate in favour of reduction, or, in his terminology, in favour of provincialism. On the whole, Rosenberg's stark opposition is not fruitful. True, in one sense, there must be compatibility. Biological organisms are made out of molecules, and they cannot do anything which molecules cannot be made to do. On the other hand, it is clear that biology, to be compatible with physics, need not be deducible from physics. Molecules, arranged in certain living patterns, may and can do things which molecules not so arranged cannot do.

There are powerful arguments in support of autonomy,[11] but in so far as they succeed, they merely tend to separate biology from physics. We are not looking for separation, but for an addition. We are looking for a world of physics which not only is compatible with the world of living matter, but also can be understood to have produced living matter, so that biology, no matter how peculiar and autonomous its laws and concepts, can be seen to be continuous with the world of physics. There are some less powerful arguments in support of provincialism which boil down to the summary conclusion that biology is an 'amalgam' of other sciences. In so far as they succeed, they tend to absorb biology into physics and chemistry. However, we are here concerned to find a viable connection between biology and physics, not with proofs that there are either two sciences or one science.

Rosenberg's remark that the biology which cannot be accommodated with physics has to be jettisoned[12] is not helpful by itself, because the crucial question is what kind of physics biology ought to be accommodated to. Do we jettison biological findings which are incompatible with Cartesian vortex physics? or with Newtonian mechanics or with quantum mechanics? Unless one is a vitalist or believes in entelechies, it goes without saying that any parts of biology which are not compatible with physics have to be discounted. The question, however, is what kind of physics? There is no way in which we can link Cartesian vortex physics with evolutionary biology, and it has seemed until quite recently that there is no way in which we can link the physics of the Second Law of Thermodynamics with the evolution of biological organisms. The question is whether there is a physics which can be linked to biology. Rosenberg, though his book was published as recently as 1985, is indeed completely silent about the new projects in physics and chemistry which have been started by the work of Ilya Prigogine[13] and Manfred Eigen[14] and which provide an interpretation of the Second Law of Thermodynamics in terms of which physics and biology can be linked – regardless of whether one thinks that, after the link, the two sciences are independent or not.

It had been noted for a long time that because of the Second Law of Thermodynamics the evolution of living matter by variational change and the behaviour of non-living matter by transformational change cannot be readily combined. This observation tended to keep biology and physics apart in some fundamental way which lent succour to the autonomists, even though, as we have just seen, they did not need it. There are quite enough arguments in favour of autonomy without the difficulties posed by the Second Law. But without some common accommodation in regard to the Second Law, the autonomy of biology continues to smack of vitalism. Non-living matter is subject to a continuous decrease of order and moves towards entropy, i.e. a state of equilibrium. The evolution of living matter, on the contrary, seemed to move in the opposite direction because it produces and maintains ordered system after ordered system and does not appear to move towards equilibrium. To my knowledge it was Erwin Schrödinger who was the first to address this problem and propose a solution. In order to maintain themselves, living systems, Schrödinger argued, suck order from their environment and maintain themselves in an ordered state by accelerating the entropy of the world they are part of. The Second Law is not violated by biology, but merely circumvented at the expense of global increase in entropy.[15] This proposal explained how existing living matter maintained itself in disequilibrium while non-living matter kept moving irreversibly towards equilibrium. But Schrödinger's proposal could not solve a crucial problem: how, given the Second Law, could some parts of non-living matter have combined themselves, as they must have done at some specific point in time, into a pattern of behaviour which results in that specific and peculiar relationship between DNA and proteins on which all subsequent evolutionary change by variation depends? How can some matter, moving irreversibly towards equilibrium, suddenly arrest its course, so to speak, and produce organised systems which cease to move towards equilibrium?

Prigogine answers that dissipative structures can arise far from equilibrium conditions. These structures arise inside a system which is moving irreversibly towards entropy; like everything else, they are dissipative. But, unlike the entire system, they are organised structures which maintain themselves in an ordered state. In classical thermodynamics, entropy was a source of disorder. In Prigogine's conception of dissipative structures, entropy can also become a source of order.[16] The non-equilibrium order principle is also likely to be important in the understanding of prebiotic evolution and of the origin of life.

> In particular, the formulation of evolution in terms of stability theory permits us to arrive at new criteria, extending, in the prebiological stage, Darwin's 'the survival of the fittest' idea. Thus the search for stability may lead ... to an increase in the quality factor, an increase in dissipation or even a local variational principle. In particular, the increase in dissipation may give rise to an evolutionary feedback that 'prepares the way' for a new instability enabling the system to evolve further. In this sense, the basic Darwinian picture of evolution is kept, but the notion of the 'fittest' is defined, in the prebiological stage, by a more subtle criterion that is no longer equivalent to that of the maximum number of offspring.[17]

The real importance of Prigogine's work lies in his extension of the theory that in biology structures are maintained in an orderly state through an influx of energy from the environment (they are 'dissipative structures' because they maintain structure while increasing the drift towards entropy) to prebiotic evolution. 'It is only recently', he writes,[18] 'that ... we begin to understand the *constructive* role played by irreversible processes in the physical world.'

One could sum it up by saying that here physics tends to be reduced to biology. But reduction is not the issue. The issue is whether there is a link between biology and physics and whether there is a physics which, in spite of the irreversible movement towards a maximum of disorder, allows for the origin of living matter and the evolution of species which maintain themselves as ordered systems. In his 'Thermodynamics of Evolution',[19] Prigogine and his fellow-workers apply the notion of dissipative structure in detail to Manfred Eigen's famous hypercycle – the model which could explain how DNA and proteins first got together in this uniquely specific relationship which made possible the subsequent evolution of living matter. This model is crucial, for it establishes the bridge between dead matter and living matter.

Unfortunately there are scientists who have jumped the gun. Instead of taking Prigogine's suggestion that his view of entropy is raising the world of inert matter to the level of living matter seriously, they are using his view of entropy in order to drag the world of living matter firmly back to the level of inert matter. In particular, they have misread the crucial passage from 'Thermodynamics of Evolution', quoted above. In this passage it is stated that 'the increase in dissipation may give rise to an evolutionary feedback that "prepares the way" for a new instability enabling the system to evolve further'. Prigogine adds that while the dissipation due to the Second Law could explain instabilities, 'Darwinian selection' is necessary to determine which of these instabilities are to be retained. In other words, entropy does not explain the evolution of living matter as such but only the emergence of variations and instabilities and mutations and errors of copying. Prigogine's theory of dissipative structures explains how it is possible, in a world which is moving irreversibly towards equilibrium, for orderly, entropy-resisting systems to emerge. But for evolution to take place, there has to be selection from the abundance of these 'errors' and instabilities, for obviously not all of them can survive and last. The meaning of evolution is *selective* retention. Neglecting Prigogine's proviso, there has emerged a whole school of thought in which entropy is simply taken as an explanation of evolution, because entropy explains the 'teleomatic drive toward configurational disorder'.[20] In this school it is supposed that the processes of evolution have nothing to do with selection processes of any kind and are driven by entropy which, for example, makes for the occurrence of errors in the replication of DNA.[21] If this project of showing that evolution is *nothing but* entropy succeeds, philosophical and ontological anxiety will be kept at the maximum described above by Russell and Sartre, for in this project, biology becomes completely absorbed into physics, where 'the view from nowhere' obtains – a view in which the observer cannot appear in any way linked to what is being observed. Looking at the world from nowhere is fascinating and even enlightening, but it carries no assurance that the findings are adequate, let alone realistic. It could be a dream or a phantasy.

The elimination of selection from the evolutionary process is not only wrong because it makes nonsense of evolution by chance mutation and selective retention, but is also in direct contradiction to what Prigogine himself has written.[22] Evolution has a dual nature. 'It deals, so to speak, both with the "vertical" phenomenon of adaptive change and with the "horizontal" phenomenon of populations, incipient species and new species.'[23] While geneticists and palaeontologists may have ignored the diversity-making (taxic) component of evolution, this school of thought, more Prigoginian than Prigogine, certainly ignores the adaptation-making component, the component of selection. In view of the general tenor of their reasoning, the disclaimer that 'in suggesting that evolution is possible only because of intrinsic changes in information systems, we do not wish to leave the impression that extrinsic factors cannot affect the evolutionary process'24 sounds lame.

Prigogine did not mean to say that evolution is nothing but entropy and follows from the Second Law of Thermodynamics. He stated explicitly that selection has to take place for evolution to occur. He merely wanted to show that in spite of the Second Law, orderly structures can appear, that their appearance is not confined to the biotic stage of irreversible process but can be found also in the prebiotic stages and that, because of the Second Law, these orderly structures have to be considered to be dissipative. In this way physics and biology become linked but there is no hint in Prigogine that the evolution of living matter by itself can be explained by the Second Law and biology reduced to physics. Once we understand that the Second Law of Thermodynamics not only does not exclude the emergence of dissipative structures but actually encourages it, we get a physics which is clearly linked to biology and very different from the physics which inspired Russell's and Sartre's ontological anxiety.

The moment in time from which we are looking at the world and knowing about it is the only moment from which looking and knowing are possible. Earlier, we – and, for that matter, other living matter – would not yet have been here, and, later, we will no longer be here. There exists, therefore, a special reason why we are looking at the world from the position from which we are looking, and that position is not nowhere, but somewhere. To be precise, it is the only possible position from which looking and knowing are possible. Thomas Nagel's argument that we *are* looking from nowhere is based on a misunderstanding of Darwinian theory.

> The Darwinian theory of natural selection … is a very partial explanation of why we are as we are. It explains the selection among those organic possibilities that have been generated, but it does not explain the possibilities themselves. It is a diachronic theory which tries to account for the particular path evolution will take through a set of possibilities under given conditions. It may explain why creatures with vision … will survive, but it does not explain how vision … [is] possible.[25]

Nagel is wrong in thinking that there is no explanation why vision is possible. On the contrary: the explanation that creatures with vision are adaptive and will survive better than creatures without vision implies that there is something there to be seen. Vision would not be adaptive unless the world were the kind of world which can be seen. The fact that vision is an advantage depends on the fact that there is light, and since there is light, it is no accident that we are as we are. There is a very good reason why, among all possibilities, light-sensitive creatures should have been selected. The Darwinian theory is more than a 'partial explanation of why we are as we are'. We are as we are because the world in which we have evolved is the world it is, and when we are viewing that world, we are doing so in direct response to the fact that it is the kind of world which has made possible the reference to itself that is involved in vision. Nagel's whole dichotomy between an objective view of the world – a view from nowhere – and a subjective view of the world – a view we have as particular subjects – is misconceived because it fails to make the most of biological evolution. It is perfectly true, as he says, that we cannot cease to be particular subjects. But the particular subjects we are have evolved and have a nearly necessary place in the world in which they have evolved. The subjective point of view must therefore be related to the objective vision. Or, to put it the other way round, the objective vision cannot be a vision from nowhere, because it cannot be had at any other time and place than the present, which is more or less coincidental with the subjective view we have as particular beings at the time and place we are in.

This conclusion has to be distinguished sharply from the position of traditional idealism which identified *esse* and *percipi*. According to this old view, the universe would be taken not to have existed before it had been ready to allow knowers to be evolved. This is patently absurd. But there is nevertheless a relation between being and knowledge. Only, it is the other way round. Berkeley believed that we could not think of the world unless we perceived it. We must turn this relationship round and state that unless the world had certain features right from the beginning, beings capable of evolving by acquiring knowledge about it would never have emerged. We can envisage an earlier and later moment in the history of the universe in which there was and will be *esse*. All we are saying is that there is a necessary relationship between *esse* and *percipi*; and that knowledge is not possible, except at a certain point of *esse*, a point which can only be reached because the forces operative in the universe have the special strengths they have. This is the very opposite of Berkeley's old idealism.

The place from which we know and think is not arbitrarily chosen. It is not an Olympian place, in an imaginary outer cosmic space. The knower does not appear as an alien. He appears because the universe in which he has evolved has certain features which have made that evolution not only possible, but well-nigh necessary. The link between the fact that we are here as a result of knowledge about the environment, i.e. because of evolution, and the fact that the universe has certain characteristics which have made for this evolution is known as the Anthropic Principle.[26]

The Anthropic Principle stands on its head the traditional question as to what kind of 'mind' is required to know the world. In its traditional form, this question cannot have an answer, because an answer would presuppose that we already knew what the world was like so that we could form a conception of the kind of mind or theory required for it. The question which follows from the Anthropic Principle asks what kind of world must exist for us to have evolved and be here. The Anthropic Principle shifts the burden of ignorance from the mind of the knower to the world to be known, and, in part, is able to answer it and remove the ignorance because we are here: if the world were significantly different, we would not be here. We know we are here and we have a lot of knowledge about the world, and the question therefore ceases to be: what kind of mind should we have? and becomes, instead, the question: what kind of world must it be, for us to be able to have knowledge of it, i.e. to have evolved by the acquisition of knowledge? The position we are in when the question is put can be located in space and time and is an identifiable place in nature. It provides a view from somewhere. Contrary to the view from nowhere which gave rise to the picture of physics, the Anthropic Principle, putting the world of physics and the world of biology together, states that we cannot discuss the universe unless it includes us. The presence of knowers is a clue to the nature of the universe which must be such that they could have evolved. Since question and answer are part of the natural world, we must view the acquisition of knowledge as a purely natural process. With this conclusion we turn to the proposal of Quine that epistemology should be naturalised and that the science of science is a science.[27]

III

It all depends, of course, on the meaning of 'naturalisation'. In Quine's proposal, it is not a question of reducing epistemology to physics and psychology so that cognitive relationships appear to be, in reality, nothing but causal relationships as studied in physics and psychology. In this view, the term 'naturalisation' would mean that epistemology is naturalised when the notion 'cognitive' is dropped and all relations are seen to be causal. Dewey, for example, believed that a philosophical treatment of epistemology is a stop-gap until science can solve all cognitive problems. Wittgenstein and Ryle thought that if epistemological problems could be solved by scientific methods, they were not philosophical to start with. Quine is thinking of something more subtle than reduction or semantic reformulation. He is suggesting that when we take cognitive relationships seriously, that is, when we realise that they are a basic and essential component of evolution and that if they had not been developed we would not be here, the purely philosophical treatment of epistemology becomes redundant. A philosophical treatment of epistemology always started with the question which norms should prevail and which norms should be set up as the criteria for valid knowledge. As we have seen, this approach is condemned to failure because any answer to this question presupposes that we already know what the world is like and what norms for knowing what it is like would be appropriate. It presupposes that we know what we are trying to know. It presupposes, for example, that we know that the world is full of dense matter or that the world is our imagination, or that the world consists of nothing but universals, and so forth. Now that we can see that cognitive relationships – the subject matter of epistemology – are a natural condition of our presence, we can study them as we study everything else and can abandon the search for the norms which make those relationships valid. They must be valid because they have produced the evolution of living matter. If they were not valid, living matter would have come to a stop long ago; or, as Simpson once put it, a monkey who habitually misjudged the distance between branches would soon be a dead monkey and leave very few offspring, if any.

Quine's description of the naturalised study of epistemology with the search for norms removed is disarmingly general:

> This human subject is accorded a certain experimentally controlled input – certain patterns of irradiation, for instance – and in the fullness of time the subject delivers as output a description of the three-dimensional external world and its history. The relation between the meager input and the torrential output is a relation that we are prompted to study for somewhat the same reasons that always prompted epistemology; namely, in order to see how evidence relates to theory and in what ways one's theory of nature transcends any available evidence.[28]

This beautiful blanket statement ought to be considered in its several parts. To start with, though Quine does not say so, he is thinking of biology – that is, of the self-referential relationship of matter to living matter, of matter to itself in a different organisation. An earthworm is living matter which refers to other matter. It embodies a great deal of information about its environment and its behaviour (its torrential output) is indeed governed by 'meager input', that is, 'meager' in relation to its behaviour. The difference is supplied by its a priori knowledge, that is, by its rudimentary nervous system. If the relationship between input and output were purely physical, and if no a priori knowledge were involved, there would be no discrepancy between input and output. The worm would have to lie still in the earth until some earth movement pushed or pulled it. The chances are that it would thus starve to death. When Quine continues to say that when we are studying the earthworm we are studying how theory transcends any available evidence, he seems to be taking the biological cognitive relationship as the model of knowledge and is not confining his attention to physico-causal relationships which never go beyond the information given. If we are thinking of physics, we are thinking of perception as energy transfer, because we are thinking of something in the world acting causally upon the organism to 'cause' perception and, in this case, one would be hard put to account for the discrepancy between input and output. Quine's terminology remains misleading because although he is mentioning the discrepancy, he is also thinking of a causal push when he is saying that there are 'patterns of irradiation' which, 'after the fullness of time, deliver an output'.

We ought to understand the expression 'meager input' correctly. In reality, there is not only no meager input but no input at all, because theory formation as well as the evolution of embodied theories (organisms) is a priori and precedes input. However, once the theory has evolved, there must be an observed initial condition to trigger the response, which is truly torrential compared to the meagerness of the initial condition. This view is compatible with Quine's argument and merely clarifies the role of the meager input and reinforces his own frequently stated view that knowledge is not built up piecemeal from sense-data or *Protokollsätze.* It is clear, however, that Quine sees the establishment of knowledge and the relation between input and output as due to a progression of physical causes, even though input and output are disproportionate. Ultimately, in his scheme, the 'irradiation' *causes* the theory about the three-dimensional world and its history. In this way, Quine remains completely beholden to physics. It would seem more realistic to discount the progression from input to output altogether and to acknowledge that knowledge about the three-dimensional world is in place a priori. In the case of embodied theories (organisms) it is inherited from naturally selected ancestors, and in the case of disembodied organisms (theories) it comes by rational selection from wild-cat proposals. The biologically established cognitive relationship (i.e. selective retention of proposals) accounts much better for the disproportion between input and output than Quine's quasi-physical progression from input to output.

When we come to the view that all epistemology falls into place as a chapter of psychology and/or natural science, we must make an important distinction. As far as the evolution of organisms on the earlier rungs of the scale are concerned, the establishment of their cognitive programme is indeed a purely natural process. They evolve as proposals put to the environment. The environment favours and thus selects those whose programme happens to arouse expectations about the environment which can be met better than those of their competitors. In this way we can explain how organisms function by self-organizing neural processes which achieve a certain end-state of interaction between the organism and its environment in a flexible and adaptive manner.[29] On this early level of evolution the Quinean suggestion that epistemology must be naturalised must succeed.

But as soon as we reach organisms capable of consciousness, there is less plain sailing. Consciousness by itself is no great impediment. There is good reason for supposing that all sorts of mammals are capable of consciousness. But such consciousness, as long as it remains non-verbal and inarticulated, does not appear an impediment to the purely natural acquisition of knowledge, nor is there any sign that its presence interferes with knowledge once it is in place, e.g., once cats have evolved. A cat is most probably capable of inchoate conscious stirrings, but there is nothing in its behaviour which interferes with the expectations aroused by its genetic programme. But once consciousness and the evolution of three-dimensional language come together (this coming together, as we have seen in the Introduction, is no accident), the picture changes. The presence of consciousness affects the 'output' in two quite different ways.

First, though the acquisition of knowledge continues to be a natural process, that natural process becomes, so to speak, hypothetical. The phenomenon of consciousness is, as was argued in Chapter 1, inchoate and unworded. For this reason – and not because it is 'inner' or private or privileged – it is ineffable. However, with the ability to formulate sentences, human beings attach labels to the phenomenon of consciousness. A labelled state of consciousness becomes expressible and can be communicated. But since the label is formulated and attached according to the conventional rules obtaining in a speech community, it does not stand in an essential or necessary relation to any state of consciousness itself. This situation puts paid to the fond hope of psychology as a science, for the hypothetical character of the labels shows that the raw evidence of psychological theory is not raw and psychological, but moulded and defined by social pressures and conventions. Psychologists have often noticed that, for example, an experience which is said to have damaging results in one social context leaves no marks of any kind in a different context. But they have hardly ever reached the ineluctable conclusion that psychology is more of a social science than a science of the mind or of consciousness.

Alternatively, I have argued in the same chapter, states of consciousness can be symbolised rather than merely labelled. While such symbolisations provide a more satisfying awareness of appropriateness, they, like worded labelling, can never be considered fully adequate to reflect or represent the state of consciousness. The symbol, it was argued, either says too little or too much, but can never get it right. There is again a very simple reason for this state of affairs which is very similar to the reason why labelling can never be more than hypothetical. To get a symbol right, one would have to have independent knowledge of what is supposed to be symbolised. But since the state of consciousness which is to be symbolised is inchoate, there cannot, by definition, be such independent knowledge. Therefore, the expression 'right' in relation to either symbol or label cannot make sense.

If the relation between the state of consciousness induced by neuronal activity and label or symbol is hypothetical, we clearly have to look in directions other than those envisaged by cognitive science to gain an appreciation and understanding of the nature of knowledge in human beings. A purely empirical study of the species of primates in question cannot suffice. One has to take into account, first of all, the intractable nature of inchoate consciousness itself. It may be inchoate and ineffable; but it is not totally plastic. It can be shifted and defined, up to a point. But beyond this point, it offers resistance. It may be quite uncertain where this point is located, but the ultimate certainty of the recalcitrance of conscious states beyond that point is inevitable. This indicates that eventually and certainly beyond that point we have to take the conventions of the speech community into consideration. And also, when it comes to symbolisation, we have to take depth psychology and the purely cultural traditions which determine the availability of symbols into account. There is nothing unnatural in these requirements. But they make it clear that the acquisition of knowledge, beyond that point, cannot be considered the product of input, no matter how great a disproportion between input and output we are prepared to envisage. The end-product of information transfer and retention cannot be assessed in terms of a relationship between the organism and its environment, but results in part from the speech and culture community of which the organism is a member. This is not so right across the board. In organisms where there is no consciousness or where consciousness is not accompanied by articulate, three-dimensional speech ability, Quine's straight naturalism is appropriate. It only ceases to be sufficient when articulable consciousness becomes involved.

At this point the reader may well sigh and conclude that with this admission the whole of the argument against sociologisation of knowledge in Chapter 3 is now redundant. But such a conclusion would be a mistake. For even with this admission, epistemology is not handed back to the philosophers who have sought to derive it from certain normative requirements – such as that it has to be reducible to sense-data or that it must be deducible from self-evident principles. Nor does the recognition of the importance of the rules of speech obtaining in any one speech community transform the rational normative requirement into the pseudo-normative requirement that a hypothetical label, to be valid, must have the consensus of the speech community one happens to be part of. For it has been shown in Chapter 4 that for a verbally formulated theory (disembodied organism) to be preferred selectively, it has to obey a number of criteria which are not arbitrarily decreed by a speech community or a cultural tradition, but which reflect quite strictly the minimal ontological features of the world – minimal in the sense that if they were not the case and that if the world had other features instead, we would not have evolved and been here to think about this matter. If these criteria are called 'norms' because they do not arise naturally and automatically, they are 'normative' only by courtesy and in a manner of speaking. The statement that one ought to love one's neighbour is certainly more normative than the statement that one ought not to jump out of aeroplanes. The latter reflects the nature of gravity while the former, arbitrarily, decrees a standard of behaviour. Nevertheless, for a complete naturalisation of epistemology to be possible, we would have to be able to observe a human being and its environment and also observe the output it produces in the form of statements about the world and then explain the former in terms of the latter or assess that the latter is accurate or adequate to the former. Where no consciousness is involved or in organisms where consciousness is present but does not play an identifiable role in the acquisition and storage of knowledge, such naturalised epistemology is all we need. But in organisms in which the process of acquisition and storage goes *via* consciousness, such assessment in terms of a relation between environment and organism is clearly not possible, so that epistemology cannot be a purely naturalistic study even though it must be granted that the traditional normative inquiry has to be eliminated from it.

Consciousness, as the foundation of knowledge, has had a deservedly bad press. When one thinks that consciousness is the one single hard piece of evidence, one can, as Cartesianism and other forms of idealism have amply shown, never get beyond the phenomenon of consciousness itself. In our century, repeated efforts have been made to circumvent consciousness as a case of double-think or as altogether unnecessary. But the one great obstacle to all forms of behaviourism is that no variety of behaviourism can explain why we have consciousness, i.e. why it has evolved and why it has been selectively retained. It has even been thought that with the acceptance of evolution, the phenomenon of consciousness as a spring of action or thought can be eliminated. But in reality, evolution has hardened the evidence that consciousness must play a crucial part. If it did not, it would not have evolved. Since it has evolved, it must be adaptive, however obliquely so. The theory of evolution has reversed the trend by which consciousness was discounted or explained away as double-think and obliges us instead to consider it an irreversible phenomenon which has to be taken seriously. As explained in Chapter 1, the most striking side-effect of the presence of consciousness is that it has pressured human beings into projecting a picture of the world which can mirror consciousness and point to its large variety of nuances and shades. Such pointing cannot be done by direct reference, because states of consciousness are ineffable and cannot be described except by hypothetically assigned labels. This picture of the world is the picture composed of symbols. When these symbolic images are formalised according to aesthetic criteria, we obtain art. When they are formalised according to logical criteria, they turn into concepts and we obtain metaphysics.[30]

'Origin of man proved. Metaphysics must flourish.'[31] As so often, Darwin was right. A. Rosenberg comments correctly[32] that though, in the first instance, the influence of Darwin's theory was one of the main causes of the disappearance of philosophical and theological speculation, in the long term Darwin was right: 'For no theory has had a greater impact on providing biological knowledge than his, and this expansion of knowledge must result in the flourishing of metaphysics.'[33] In so far as evolution provided a naturalistic explanation of the origin of man, further thoughts of metaphysics indeed became redundant. But in the long term, a naturalistic explanation, more than any other, must account for the non-naturalistic habits of man, for their persistence and their ubiquity. It is not entirely clear what Darwin precisely meant by 'metaphysics'. But we can suppose that he wanted to indicate that since human beings had evolved by natural selection, every one of their faculties, including the faculty of ineffable subjective consciousness, must be adaptive. And since the presence of consciousness creates the symbols which, when conceptualised, are metaphysics, metaphysics will flourish now that 'Origin of man proved'. As long as one had no clear idea why we had consciousness and where it came from, it was possible to consider consciousness as nothing but a causal factor in behaviour and all or any *thought* about it, productive of error. Hence the temptation to discount both symbols and their conceptualisation – the only way in which subjective consciousness could make itself manifest – as mere metaphysics, remained strong and urgent. Now that we know that consciousness is a product of evolution, these temptations can and have to be resisted.

The science of science is, itself, not a complete science. Even though such norms as we use in science are ontologically necessary and therefore not really extraneous to science, the judgement that they are not and that they are necessary is a philosophical, not a scientific, judgement. Whichever way we turn, the philosopher is left standing as a kind of overseer of knowledge, alone able to determine its quality. His role is limited, but it is inevitable. The word 'overseer' has an ugly sound and reminds one of the age of slavery. With this unattractive connotation in mind, it was used as a description of a certain kind of non-conversational philosophy by Rorty in order to discredit the non-conversational philosopher's role and to bring it into disrepute. Rorty wants us to believe that these unavoidable philosophers are like dictators with a 'know-all' attitude. But since science cannot be completely naturalised because it cannot be understood without some philosophical interpretation, extra-scientific and meta-cognitive judgements are needed to assess the degree of correspondence of knowledge with reality. Such meta-cognitive judgements must be philosophical judgements. Willy-nilly, the philosopher remains in a role of 'overseer'. Rorty's objections must therefore be brushed aside, all the more so as his own arguments against the overseeing role of philosophy are much less subtle and even more vulnerable than those of Quine. For Rorty not only supports the autonomy of science and rejects the need for normative evaluations on the grounds advanced by Quine, but also misunderstands it by insisting that it does no more than produce a 'double' of what there is. Since such a description of scientific knowledge bears no resemblance to what we understand by science, it ends up by making nonsense of Quine's view that science is its own master and that epistemology is natural. One might even consider it as a *reductio ad absurdum* of the contention that philosophers should not be overseers. People who can persuade themselves that science is merely a duplication of reality are living proof that they stand very much in need of a philosophical overseer![34]

IV

Since biology itself furnishes the norms, it seemed a philosophical consequence of biology that epistemology can be naturalised because it need no longer search for the norms which make knowledge true. However, the Quinean project of a naturalised epistemology has come up against both sociology and metaphysics, and it is therefore misleading to speak of naturalisation. It has instead become customary to refer to the cognitive consequences of biology as evolutionary epistemology – that is, as a theory of knowledge which solves many or most traditional problems of epistemology with the help of the theory of evolution. There is no consensus about and not even a clear understanding of what is meant by evolutionary epistemology. The subject is frequently mentioned, but none of the many references to it have much in common. The well-known book by Rupert Riedl[35] presents a vast amount of ethological and biological evidence but very little philosophical analysis, and the book by Gerhard Vollmer,[36] which links philosophical analysis to biological evidence, is, as will be discussed later, faint-hearted because it confines evolutionary epistemology to the pre-scientific part of knowledge. Stephen Toulmin[37] and David Hull,[38] by contrast, confine evolutionary epistemology to the development of science. Let me therefore recapitulate and summarise the basic contentions of philosophical Darwinism as they have been developed in this book. As a body, these contentions are more comprehensive and more daring than all previous contributions to the subject because, in one continuous sweep, they link the evolution of organisms to the evolution of scientific knowledge.

1. Both embodied theories and disembodied organisms are theories about their environment. They store hypothetical knowledge about their environment in the form of information about regularities. The information is transferred to them by natural or artificial selection and is not stored in their minds.

2. The stored information is a form of self-reference of one part of the world (living matter) to the rest of the world. Such self-reference is only possible at a certain moment in the time of the evolution of the universe. The referring or reflecting mirror has evolved in that universe and is part of what it reflects.

3. Information is transferred from one part of the world to the other by a double process in which there is an abundance of proposals and selective retention. The abundance of proposals is, theoretically, by chance; and the selective retention is the necessary effect of exposure of that abundance to the environment. Very few proposals survive; but those that do survive are adaptive and verisimilitudinously true.

4. In every theory and in every organism, this information is in place before theory or organism is exposed to the world. The information is not due to a

learning process or to instruction of the organism or theory by the world.

5. The theory defines an organism's or theory's *Umwelt*, which is that part of the environment the theory directly refers to. The theory constitutes the *Umwelt*, and the latter would not exist without the former. The *Umwelt* is defined by the expectations aroused by the theory. The theory defines the initial condition which, when met, triggers a response or prognosis. The theory refers to the *Umwelt* because the expectations it arouses constitute the objects or events which act as initial conditions. The problem of reference – of how concepts or words hook into the world – disappears because it has become clear that the objects and events in the world only appear in so far as they are constituted by the theories and organisms.

6. The biological model of knowledge does not distinguish sharply between organisms as embodied theories and theories as disembodied organisms. Evolution by knowledge acquisition is continuous from the 'amoeba to Einstein'.

7. However, there is a difference between the manner in which organisms are selected naturally and the way in which theories are selected rationally. As far as theories are concerned, selection is by a rational decision to give preference to those theories which explain more. Although this decision is normative, it is not arbitrary. It is rational because it reflects a basic feature of the universe. The universe consists of sub-systems, but all sub-systems must be compatible with each other. Therefore the progressive removal of inconsistencies by preference for those theories which explain most is a progressive approximation to truth by correspondence.

These seven propositions can readily be divided into two groups of ideas, but not into two groups of propositions because, advisedly, both groups of ideas occur in all seven propositions. One group of ideas refers to the biological evolution of organisms from protozoa to *homo sapiens,* and the other to the evolution of knowledge as a body of linguistically formulated proposals, an evolution which begins with the emergence of three-dimensional language. The two groups between them and the seven propositions together form the core of philosophical Darwinism, the claims of which are more specific than any or all of the many claims which have been made in the name of evolutionary epistemology.

As to the ideas of the first group, there has been little to quarrel with. We have come a long way since the days of Lord Monboddo in the eighteenth century, who believed that pre-human animals learn by experience and are waiting to be instructed by the world. He believed that the reason why they never became as clever as human beings was that they did not live long enough to learn properly! Today we all accept that organisms do not learn by experience but inherit their knowledge; that that knowledge is in the genetic programme through natural selection and, therefore, ready a priori; and that both learning by conditioning (Skinner) and by imprinting

(Lorenz) are very limited modifications not of the inherited programme but of the environment so that the organism is compelled to be fooled just because his programme is so fixed. The only new element in this group of ideas is the contention that evolution is a knowledge-acquisition process. But even this contention concerns philosophy more than biology. For philosophers rather than biologists may question whether the concept 'knowledge' can be extended sufficiently to cover biological adaptation.

The real and formidable doubts about the feasibility of philosophical Darwinism concern the ideas in the second group – the contention that the evolution of knowledge is continuous from the amoeba to Einstein and that theories are disembodied organisms because they evolve, like organisms, by chance mutation and survive by selective retention. Since we are going on the assumption that the evolution of organisms is a process of knowledge acquisition and that that evolution takes place by chance mutation and selective retention, the continuity thesis follows because it states no more and no less than that knowledge develops by chance mutation and selective retention and that this process of acquisition continues beyond the threshold of articulable consciousness. Theories, therefore, are disembodied organisms. Although this continuity thesis is the direct consequence of the feasibility of regarding theories as disembodied organisms which evolve by chance mutation and selective retention, I will refer to the whole complex of ideas as the continuity thesis. Discussion of the feasibility of continuity, however, must concentrate on whether theories can be regarded as disembodied organisms. The continuity thesis follows if it is feasible to regard them as disembodied organisms. There is nothing novel in the contention that biological evolution is a process of knowledge acquisition, and there is no serious opposition to biological evolution. The only slightly novel part of this view is the idea that evolution is a cognitive process. But this is a matter of semantics, rather than of philosophy. The rest – the contention that biological evolution results in organisms whose cognitive apparatus is adaptive and adequate for survival – is not controversial. This possibility was first mooted by Ernst Mach when he pointed out that our subjective cognitive apparatus could not be all that subjective because it was biologically conditioned.[39] It was fully elaborated by Konrad Lorenz.[40]

But there is an interesting irony here. Both pioneers and elaborators of this view were philosophers who believed that our knowledge is built up piecemeal, by induction, from sense-experiences, subjective perceptions, sense-data or whatever. They were therefore vitally interested in showing that our sense-experiences must be veridical, because mediated by biologically evolved and adapted sense-organs. Once they had satisfied themselves on the veracity of our senses, they believed that the rest could safely take care of itself. Since they believed in induction from sense-experiences, they held that evolutionary epistemology has done its job when it has established that our sense-organs are adaptive and therefore reliable. Mach, especially, was looking for the foundations of secure, scientific knowledge. Where Husserl claimed to have found them in a transcendental consciousness – i.e. in the kind of consciousness which remains when the world we are conscious of is 'bracketed' and belief in it suspended – Mach was looking for these foundations in those subjective experiences which were part of the cognitive apparatus evolution had produced by natural selection. The whole early interest in evolutionary epistemology was governed by the search for the foundations of knowledge.

The focus of interest shifted radically when it came to be seen that our sense-experiences, at most, play a negative, falsifying role in the construction of knowledge. Once this was understood, evolutionary epistemology had to change. The main emphasis now had to be not on the adaptiveness of our senses and the veracity of the knowledge they mediated, but on the continuity of the evolutionary process and on the continuity of selection. Thus the idea that theories are disembodied organisms moved into the centre, and the idea that sense-experience is veridical because the senses have evolved by natural selection, though important, ceased to be the sole concern. However, the contention that theories are disembodied organisms and that the growth of knowledge beyond the threshold of consciousness is continuous with the growth of knowledge below that threshold is novel and has met with serious opposition.

Criticisms of the continuity thesis in this strong form in which the continuity between biological evolution and the evolution of scientific thought is not a mere analogy or a metaphor have been direct and explicit. But they can be rebutted. Let us start with the criticism raised by Gerhard Vollmer, the author of the first full-length book on evolutionary epistemology. He bases his rejection of the continuity thesis, on the contrary, on the very strength of the argument in favour of the adaptiveness of our sense-organs and their cognitive apparatus. In his *Was können wir wissen?*[41] Gerhard Vollmer writes that we can think of the impact of the world on the human organism as energy transfer on which we can depend because of natural selection. Thus we can account for knowledge as a causally generated phenomenon and dismiss Hume's scepticism that there is no relying on causality because all we have is regular sequences. The very fact, Vollmer argues, that we have perceptions entitles us to think that that perception is causally produced through energy transfer from an object to our retina and from there into the appropriate part of the brain. However, and this is crucial, Vollmer softened this hard evolutionism in regard to perception by the proviso that the knowledge so generated is only knowledge about the mesocosm, that is, about the world of middle-sized objects and medium speeds. Below, there lies the microcosm of quantum phenomena, and above it a macrocosm of nebulae and red-shifts which we have knowledge of, albeit not by this kind of energy transfer. With this qualification, Vollmer denies continuity and thinks of knowledge beyond the mesocosm as knowledge acquired in a different way. He is quite explicit on this matter and believes that evolutionary epistemology can either follow in the footsteps of Lorenz, study the evolution of cognitive systems and account for knowledge of the mesocosm; or plunge into giddy depths by following Popper's interest in the evolution of scientific knowledge.[42]

Vollmer, however, cannot be right. First, both microcosmic and macrocosmic events, to be part of knowledge at all, must be mediated through events in cloud-chambers and telescopes which we do observe with our senses. This shows that mesocosm and both microcosm and macrocosm must stand in an intelligible relationship to one another. 'The fact remains', writes Max Delbrück,[43] 'that the physical sciences represent the actual or potential experiences and observations of individuals, in however abstract a form, and as such are as psychic as any emotion or sensation. Both the blue of a summer sky and the 4,400 Å wavelength of its light refer to experiential acts, differing principally in the affective components accompanying these acts and in their expressions.'[44]

Second, Vollmer's initial account of learning by energy transfer leaves something out and is not consistent, even for quite primitive organisms, with the process of natural selection. The sense-organs, or, in primitive organisms, the sensitive cells, precede, as do so many proposals, contact with the outside world. The outside world selects some of the proposals; and the selected proposals then have, because of selection, the ability to perceive initial conditions as defined by the organism-theory. In such perception there may well be energy transfer, as Vollmer says. But it is not constitutive of the information stored by the organism-theory, nor does its functioning act as guarantee that the organism perceives correctly what is actually there. The correctness of the perception of the initial condition which then triggers the response comes from the fact that the organism-theory has evolved by natural selection and would not be there if its expectations (i.e. the theory it embodies) would not be at least more or less true or, in our terminology, verisimilitudinous. If organisms get it right, it is not by energy transfer. They get it right because there has been an abundance of proposals from which those that get it right have been selected and retained. This method of getting it right can also operate for disembodied organisms about matters which are not directly experienced through energy transfer and thus applies also to theories about the microcosm and the macrocosm.

In another place[45] Vollmer goes on the assumption that we first must have a conception or image of a world and can formulate a theory about it only after we have such an image. On this assumption it is indeed correct to say that it is impossible to formulate theories about a world we cannot imagine conceptually – such as microcosm and macrocosm. The absence of concepts of these worlds derives from the fact that our sense-organs are unable to experience microcosmically and macrocosmically, because they are adapted to the world they have evolved in, i.e. to the mesocosm. But Vollmer's assumption is mistaken. No knowledge, even the knowledge that is embodied in protozoic organisms, is generated *after* the organism has taken imagistic or conceptual cognisance of the world. Similarly, theories are generated about both microcosm and macrocosm without prior images or conceptions of microcosm and macrocosm. Once the theory is there, it can be tested by the use of apparatus which mediates between the events postulated by the theory and our senses. Some of these theories remain science fiction; some, like quantum mechanics, can be tested. But as the case of quantum mechanics has notoriously shown, we can have perfectly good theories without being able to form a 'conception' of the quantum world the theory is about.[46]

Vollmer concludes that, therefore, the formation of quantum mechanics is beyond the reach of an evolutionary epistemology. But his conclusion does not follow. In principle the formulation of quantum mechanics theories is no different from the formulation of any theory about the mesocosm. For all of them, even theories about the mesocosm, go beyond the information given. It makes therefore no difference whether we can, once the theory is generated, form a conception of the world it describes and refers to or not. Vollmer's difficulty with knowledge of the world outside the mesocosm we are adapted to stems from the fact that in his heart of hearts he subscribes to some form of positivism and cannot get himself to suppose that all knowledge is a priori. In his thinking, one must have an image of a world first and can proceed to theories about it only once such an image is established. With such thinking he has to confine application of evolutionary epistemology to the mesocosm. The real trouble with this idea is, as was pointed out above, that it is mistaken even as far as knowledge about the mesocosm is concerned.

The next criticism comes from Paul Thagard. While genetic mutations are blind, he argues, mutations of thought are not and innovative theories are not.[47] Therefore new theories have the appearance of being goal-directed. If one probes beneath the surface, this turns out not to be true. If one surveys at any one time and any one place all the theories that are competing for retention, one will find that they are as random as genetic mutations. One has to take astrology, necromancy, faith-healing, colour therapy, Armageddon expectations, etc., etc. into account in order to form a proper picture of the randomness of all the theories which are offering themselves for selection. By looking at successful theories in retrospect, Thagard creates the impression that they were indeed premeditated and designed to make things better. But if one looks at biological evolution in retrospect, one could also gain the impression that mutations were not blind, but took place 'in order' to produce gills and bronchial tubes or whatever. Thagard's argument does not depend on an examination of mutations of theories, but on the fact that he is looking backwards and is examining scientific theories in retrospect and biological organisms in advance of their appearance. At the same time, while genetic mutations are random, there are myriads that are so unrelated to the environment, and therefore so irrelevant, that they do not even produce a living organism which can present itself for selection. We also have to take into account another factor which limits blindness on the organic level.

> The range of possible variation is limited by the constitution of the organism's genotype and by the laws of molecular biology and the range of mutations which a gene can undergo is restricted or determined by the gene's structure. Hence the mutational repertoire of a gene pool is restricted or determined by the evolutionary history of the species, just as the repertoire of new ideas in science is restricted by tradition or the world picture.[48]

The constraints on biological evolution which are due to the fact that selection always works on last year's models is also known as phylogenetic or ontogenetic inertia. Mutations that enter the race are, therefore, not as random as pure statistics expects them to be. A fish born without gills does not really play a part in evolution. Nor does number mysticism play a part in the evolution of the mathematics which led to the differential calculus. 'Blindness', in other words, is not an absolute, not in genetics and not in the generation of new thoughts.

Anthony O'Hear has argued forcefully[49] that no conclusions of epistemological interest can be derived from the fact that when organisms avoid predators, their perceptual faculties must be part of their success. He goes on to say that we cannot relate a frog's 'beliefs' about the world to its extremely limited perceptual faculties which go no further than an ability to spot, as an initial condition, black moving things of a certain size. The point, however, is that the frog's 'beliefs' are precisely concentric with its perceptual ability and that this ability consists in its expectations and these expectations, when met, define its *Umwelt*. The frog, in other words, is a theory about the part of the world which is its *Umwelt*. There are no epistemological conclusions to be reached. The existence of the frog is identical with a true and adaptive theory about its *Umwelt*. O'Hear is right in thinking that nothing much about the world at large can be learnt from the frog's ability to spot a moving black object. What can be learnt, however, is philosophically significant. We can learn from the example of the frog that every single organism is a theory about an environment and that the expectations that theory embodies define an *Umwelt*. We can then further conclude that since all *Umwelten* are specific to their organisms, we must wait for the evolution of consciousness and three-dimensional language to produce theories which will transcend all single *Umwelten*.

Next we turn to the criticism that selection ceases to be natural when we come to the selection of theories and thoughts.[50] This criticism is valid. In the evolution of thought, the theories which are retained are not selected automatically. This matter has already been discussed above, in Chapter 4.

The continuity thesis can nevertheless be defended against this criticism, because the criteria used for selection of theories are not arbitrarily chosen, nor do they depend on the consensus of a community. If they did depend on a consensus of a community, they would still be arbitrary or, at least, relative to a given community. The criteria, as was explained above, are dictated by the minimal ontology which is known to obtain, because if it did not, no biological evolution would have taken place. So, while in the evolution of theories the selection is not automatic in the sense in which it is natural and automatic in biological evolution, it is made to proceed, for very compelling reasons, according to the same criteria.

Next comes the criticism which concerns the question of progress.[51] The concept of progress is a very thorny concept because there cannot be agreement as to the goal which, when reached or approximated to, would constitute progress. In biological evolution there is clearly no obvious progress, even though, say, primates are more complex than worms, and worms more complex than bacteria. But even without a concept of progress, it is clear that the whole drift of evolution is a drift away from a primitive state and that it is no accident that mammals evolved after and not before bacteria. It is therefore simply not correct to say that there is no progress in biological evolution, even though the word 'progression' may be more suitable than the word 'progress'. There certainly is a progression *away* from a primitive state. In the evolution of thought, the notion of progress is more marked. Since all conceivable *Umwelten* must be compatible with one another, it is imperative for knowledge to be improved by the progressive removal of inconsistencies. This means that those theories which allow us to explain more events or phenomena are to be given absolute preference. And this preference provides the drive towards a progress which consists in fewer and fewer theories which explain more and more.

The final criticism concerns the notion of adaptation.[52] This criticism, which is based on the view that 'adaptation' in organisms and 'truth' in theories are not synonyms, is justified. However, it can be met if we are prepared to adjust our concept of 'truth'. As we have seen repeatedly, it makes no sense to use the term 'truth' when there is a correspondence between, say, a sentence and the facts. A sentence is not the sort of entity which can 'correspond' to anything other than another sentence. An adaptation, on the other hand, is not a perfect 'fit' to the environment. In many cases we are entitled to call an organism 'adapted' not so much because it is a good fit to its environment, but because there happen to be few competitors around. We can therefore try to bring the concept of adaptation and the concept of truth very close together when we accept that they are not absolutes. Any degree of adaptation indicates that the information about the environment in question is not wholly incorrect. And, similarly, the greater the consistency of a theory with other theories, the greater its approximation to the true conditions of the environment – i.e. the greater its verisimilitude.

V

For positivists, there had never been a special problem of the growth of science. Knowledge had been thought to produce its own history. Anybody who had read Bacon's *Instauratio Magna* or *The Advancement of Learning* would know how to write it. It would be the history of the accumulation of correct observations. If one goes, for example, to Thomas Sprat's *History of the Royal Society* of 1667, one finds that this is how he saw it towards the end of the seventeenth century. He equated the destruction of rhetoric with the growth of knowledge and showed that positive observations add up to theory and the growth of such theories to the history of knowledge. The point is that in those early days knowledge was believed to be based on positive observations and the history of knowledge simply resulted from the accumulation of these observations. But when positivism started to decline because the Duhem–Quine thesis had shown that there is no clear-cut distinction between observation and theory, it became clear that a special explanation of the growth and development of science and knowledge was needed. When observations came to be seen as playing a very subsidiary role in theory formation, change of theories required a special explanation which could not be furnished in terms of observations made.

The real merit of the continuity thesis is that it contains an explanation of the dynamics of theories. Theories are replaced, the continuity thesis implies, when a more adaptive or verisimilitudinous theory is found. A theory is more adaptive or verisimilitudinous when it explains not only what earlier theories explained but also something else. The dynamics are controlled by preference for theories with the greatest explanatory power. Such preference is an artificial selection of theories. Though artificial, the selection is not arbitrary, let alone subjective, because it is carried out according to the requirements of the minimal ontology.

The earliest attempt at an explanation of what the dynamics of theories are goes back to Hegel. Hegel was the first to explain these dynamics in their own terms: he did not rely on belief in a pre-ordained succession of theories and beliefs and did not think that beliefs change in obedience to a developmental law. Hegel had no knowledge of natural science, but quite successfully elaborated an account of the evolution of what we nowadays would describe as world-views. Hegel suggested that, since no one set of beliefs can do full justice to the circumstances it seeks to explain, it is essentially unstable and there always remain nagging doubts as to possible alternatives. Eventually these doubts come to the fore and generate a belief system which contradicts its predecessor. Then we get again the same deficiency, the same doubts and the same reversal. This is a perfectly plausible model of evolution and is particularly suitable for idealists who believe that there is no real world other than the one they are thinking about. However, the Hegelian explanation of theory dynamics is strictly for idealists and for idealists only, because it contains no reference to the world the theories purport to be about. It seeks the explanation of change exclusively in the nature of thinking. For Hegel, the history of thought would have been exactly the same even if some or all of the theories involved had been false in terms of the world they were about.

The first really important and specific step towards an explanation of the dynamics of theories was taken in 1934 by Karl Popper in his *Logik der Forschung*. Popper, to begin with, was not interested in finding secure foundations on which knowledge could be built. He demonstrated, on the contrary, that knowledge is hypothetical and not in need of foundations. He argued that the formation of theories, their competition with one another and the preference for some theories are an evolutionary process which continues according to the principles of biological evolution – by chance mutation and selective retention – and so provided a thoroughly non-Hegelian explanation of the dynamics of theories:

> According to my proposal, what characterises the empirical method is its manner of exposing to falsification, in every conceivable way, the system to be tested. Its aim is not to save the lives of untenable systems but, on the contrary, to select the one which is by comparison the fittest, by exposing them all to the fiercest struggle for survival.[53] How and why do we accept one theory in preference to others? The preference is certainly not due to anything like an experiential justification of statements comprising the theory; it is not due to a logical reduction of the theory to experience. We choose the theory which best holds its own in competition with other theories; the one which, by natural selection, proves itself the fittest to survive. This will be the one which has stood up to the severest tests, but the one which is also testable in the most rigorous way.[54]

This crucial text calls for a number of comments. There is, first, the striking Darwinian terminology. Second, there is an explicit denial that theories are to be preferred according to the security of any observational foundations on which they are alleged to rest. Third, there is competition among theories and selection according to a criterion of preference. And fourth, the preference criterion is 'falsification': theories which are more rigorously testable (= falsifiable) are to be preferred to theories which are less so; and theories which are falsified outright are to be deleted. In short, the competition is seen to be resolved by straight falsification. In this initial form, Popper's Darwinism of the dynamics of theories raised as many problems as it solved. The notion of 'falsification', though logically decisive (one can never *verify* a universal proposition; but one single experience to the contrary *falsifies* it), is not practical. For any theory can be salvaged by suitable changes in its context, such as *ad hoc* theories. Moreover, if the relation between theory and observation is negative rather than positive, the old problem as to how we determine which observation is relevant is not solved. A negating observation is known to be such only because it is referred to in the theory. A fact truly negating a theory can hardly show up, unless the world is seen in the searchlight of that same theory and in that light only those facts are referred to by the theory which do not negate it. Facts, we hardly need to remind ourselves, are theory-laden. Hence it is more difficult to falsify a theory than pure logic would lead one to expect. A truly negating observation is one which is not really relevant to the theory it is supposed to negate. It has indeed been very difficult to find proper examples where a theory has been falsified by observations and dropped as a result of such falsification, and for this reason the theory that theories develop by falsification and the emergence of a new theory is not a good account of the history of knowledge.

Popper also showed that theories – regardless of how they are related to observations – are not simply additive and do not follow logically upon one another. On the contrary, they stand in a very peculiar and almost mysterious relation to one another. Often they cannot be reduced to one another, and in many cases the superseded earlier ones cannot be deduced from the later ones. Thus one cannot write a history of the growth of knowledge by showing how Einstein supplanted Newton because Newton's theory can be subsumed under Einstein's theory. It was one of Popper's great contributions to have shown that the relation between, say, Newton and Einstein is neither additive nor negative. Einstein, he showed, explains everything Newton had explained, even though Newton's theory could not be deduced from Einstein's theory. The one supplanted the other without falsifying it.

This weakness in regard to falsification in Popper's initial version of the continuity thesis of philosophical Darwinism and of theory dynamics was exposed and exploited by Thomas Kuhn in 1962 in his *The Structure of Scientific Revolutions.* Kuhn not only addressed the problem which arises for the dynamics of theories when falsification, because of the rich possibilities of explaining it away and of salvaging theories by *ad hoc* hypotheses, has to be discounted as a moving factor, but also dealt with the fact that in the history of science theory systems supplant one another even though the new system has not falsified the old system:

> No process yet disclosed by the historical study of scientific development at all resembles the methodological stereotype of falsification by direct comparison with nature.[55]

Kuhn's book achieved enormous fame and influence because it provided a different account of theory dynamics. Kuhn used Popper's discovery that the relation between major theories is not additive and that major theories can neither be reduced to one another nor deduced from one another. But the book contained a vital flaw because it invoked the history of science as the final evidence:

> History, if viewed as a repository for more than anecdote or chronology, could produce a decisive transformation in the image of science by which we are possessed.[56]

The only thing one can learn from logic, as Popper had indeed shown before Kuhn, was that major theories could not be deduced from one another or reduced to each other and were in fact, in an important sense, incompatible with one another, even though, taken singly, some of their predictions failed to be falsified. The lesson which history did show was that there was room for a new term to describe those major theories which stood in this peculiar non-relationship to one another. Kuhn provided the new term and called these major theories or models 'paradigms' (also known as 'maxi-theories', 'disciplinary matrices' and 'regnant major theories') in order to show that, though explanatory, they did not stand in a relation of deducibility or reducibility to one another. But this is where he ought to have left the matter. The history which would show in which relation these paradigms did stand to one another would have to be written before it could be studied and before one could learn from it.[57] But to be written, one would need, as Joseph Agassi demonstrated in 1963 – the year after the appearance of Kuhn's book – a theory as to how these paradigms were related to one another. In other words, Agassi showed convincingly that the history one would write depended on the view one took of that relationship.[58] Thus a theory about the relationship of paradigms to one another cannot be derived from history because the history one would write would depend on one's view of the theory about the relationship of paradigms to one another. Such a theory would have to *precede* the writing of the history of knowledge and could, therefore, not itself be learnt from the history of knowledge as Kuhn had suggested. In all cases, a philosophy of science must precede a history of science because the alleged empirical record is a construction and does not lie ready, waiting to be inspected by positivists. The dependence of the history of science on a philosophy of science is still not appreciated, even by such self-styled 'post-positivists' as Arthur Donovan, Larry Laudan and Rachel Laudan, who write[59] that the claims of their own 'historical school' 'have not yet been … systematically tested against the empirical record'.

Kuhn has never confessed to this major difficulty with his evidence.[60] Instead he jumped to the conclusion that history teaches not only that there were such paradigms, but also that they stood in an irrational relationship to one another. They come and go, he maintained, because scientists die and new scientists succeed them. His theory of the development of science may stand on its own two feet, but it cannot stand on the historical feet Kuhn claims to have provided. Not only does history not provide proof, but one can also show that many paradigms stand in a perfectly rational relationship to one another, albeit not in a relation of deducibility or reducibility. Kuhn omitted consideration of the possibility that theory or paradigm preference might be guided by such rational criteria as degrees of explanatory power. Instead he turned towards sociological explanations of paradigm changes and insisted that apart from cognitively based dissatisfaction with an old paradigm, the choice of a new paradigm was due to sociological rather than cognitive or epistemic pressures and to population statistics, such as the mortality rate of scientists.[61]

It also follows from Kuhn's view of paradigms that the phenomena defined or referred to by them are incommensurable with the phenomena defined by other paradigms. In one sense, this corollary is circular. If one takes it that a term is 'defined' by the paradigm, it follows that if the term is defined by a different paradigm, it must have a different definition or meaning. The notion of meaning invariance regardless of paradigms can make no sense, and terms defined differently by different paradigms must be incommensurable, for 'commensurability' would mean that they are defined by the same paradigm. But in another, more important sense, considerations advanced by H. Field[62] show that such alleged incommensurability is hardly ever absolute. And finally one ought to realise that Feyerabend reduces the notion of incommensurability *ad absurdum*, because he bases the view that phenomena are incommensurable on the fact that it is impossible to compare a theory with a phenomenon. Nobody would claim that one can compare a general theory with a particular phenomenon unless the phenomenon itself is deduced from the theory – in which case all comparison is superfluous. In order to establish the doctrine of meaning variance and of incommensurability, one would have to demonstrate that it is impossible to compare theory with theory. But such a demonstration must fail, because one can compare theory with theory in terms of explanatory power.

There the matter stood until there appeared in 1972 the first volume of Stephen Toulmin's *Human Understanding*[63] and in 1988, David Hull's *Science as a Process.*[64] Both Toulmin and Hull have a lot in common with Kuhn. Both think that the survival of a theory or paradigm depends on the consensus of a community and that that consensus can and is given for non-cognitive reasons. In the wake of this argument there has since developed a whole body of opinion about the sociology of knowledge which seeks to explain the emergence and acceptance of theories in terms of ideological resonances. One cannot blame Kuhn for these developments. But Kuhn's non-evolutionary theory, which makes the community rather than nature responsible for selection, has gone a long way to countenance and reinforce the lure of sociology discussed in Chapter 3.

Both Toulmin and Hull, however, try to Darwinise Kuhn and attempt a more specifically biology-oriented theory of the dynamics of theories and the evolution of knowledge. Toulmin isolates four features of biological evolution, each of which, he argues, has a counterpart in conceptual evolution. In biological evolution there are, first, discrete species. Second, there is a process of competition and natural selection by which occasionally advantageous novelties become established. Third, this process gives rise to new species provided the environment favours the novelties. Fourth, the novelties are perpetuated if and only if they are sufficiently well adapted not only to the environment but also to other species. Conceptual development, he argues, runs parallel. First, there are discrete disciplines and evolution has to explain not only their temporary coherence but also the changes by which they are superseded. Second, there is a continual emergence of further novelties. If they are absorbed, there is continuity and coherence; if they survive in their novelty, the discipline is transformed. Third, for this treatment of novelties to take place, there must be a forum for competition, that is, established institutions in which informed scientists meet and defend their disciplines. Fourth, in this forum those novelties which best meet the demands of the local intellectual environment will become accredited.

Hull's evolutionary theory of the growth of knowledge uses different analogies, but it is not substantially different. In the growth of knowledge, Hull says, theories are like replicators in biology. Genes replicate and build other organisms which Hull compares to scientists, whom he calls 'interactors'. Replicated theories – none of which are absolutely true copies, so that there is a choice for selection – will be successful to the extent to which their interactors succeed in persuading other scientists to accept them. The competititon between theories is taking place inside communities of scientists.

In these models of the evolution of knowledge the competition between theories is reduced to a competition for the attention and favour of scientists. Neither Toulmin nor Hull has a clear way of distinguishing between the success of a theory by using, for example, a nerve gas which persuades; and success by an increase in rationality and explanatory power, a success which would eventually indicate the greater verisimilitude of the surviving theory. In both models the competition is for social approval, and has nothing much to do with selection on account of truth and verisimilitude, let alone with selection on account of correspondence or adaptiveness to an environment.

Less specific than Hegel and Darwinian only in the sense that there is selection of theories, Toulmin's and Hull's attempts are so vague that they explain little more than that there are changes in thinking. Both theories incline heavily towards the Kuhnian view that the ultimate determinant of theory acceptance is the consensus of a community. Their attempt to provide an evolutionary explanation for the selection of theories in place of Kuhn's conclusion that theory preference is haphazard and arbitrary depends entirely on the idea that the selecting is done by a community of scientists. There is no attempt to state the cognitive or epistemic criteria which scientists may or may not use in espousing theories. It is impossible to define the precise difference between a Kuhnian view of the reasons for the survival of a theory and the views of Toulmin and Hull. The only difference is that both Toulmin and Hull give a somewhat more evolutionary account of the non-cognitive factors which determine paradigm or theory changes.

Popper himself took up the problem again in his *Objective Knowledge* of 1972,[65] where he offered a general theory of the growth of knowledge on evolutionary lines. To start with, there is now no insistence on simple falsification as the motor. Popper here points out that ' "mutations" may be interpreted as more or less accidental trial and error gambits and "natural selection" as one way of controlling them by error elimination'.[66] This comes very close to saying that organisms are theories which solve problems and that theories are like disembodied organisms. The common factors are problem solutions and error eliminations. The feature which is not common is the fact that in biological evolution 'errors' are eliminated by the non-survival of the organisms; whereas in the evolution of knowledge, errors can be eliminated by the abandonment of theories: 'our hypotheses die in our stead'.

The present view has added very little to the Popperian schema and its continuity thesis. The only significant amendment is the contention that theories are to be preferred according to their explanatory power. There is nothing wayward in this amendment, for at every moment when there is a problem, the problem which essentially remains is the problem of insufficient consistency. In any one solution, there always remains a problem to be solved. And this is the problem that any one solution of a problem is not entirely consistent with all the other solutions of other problems. Since all of us, embodied theories as well as disembodied organisms, are inhabiting the same universe, any remaining inconsistency presents a problem to be solved. For this reason it is possible and necessary to specify the nature of the problem which always remains to be solved. This amendment, though not explicitly stated in Popper's schema, is compatible with it. There is a compelling reason for the amendment. As long as we think merely of 'problem solving', we are not saying anything other than that there are *analogies* between biological evolution and the growth of knowledge. In order to go beyond analogies and pin-point the continuity, we have to indicate what the common substantial – as opposed to the purely formal – factor in biological evolution and the growth of knowledge is. In the growth of knowledge, proposed solutions are evaluated according to their explanatory power. This decision is not arbitrary, but is dictated by the minimal ontology we know of. If there were inconsistencies in the world, if at least some *Umwelten* were incompatible with one another, we would not be here. Since we *are* here, there cannot be inconsistencies, and of any two theories which are inconsistent with one another at least one must be eliminated because it must be false. Hence, the removal of theories which are inconsistent, and the growth of theories which are consistent with one another, are steps in the direction of increasing verisimilitude.

VI

Many philosophers have levelled a very different kind of criticism against evolutionary epistemology and, by implication, against philosophical Darwinism. They claim that one cannot solve philosophical problems by taking evolution seriously. 'What is wrong with evolutionary epistemology', Hilary Putnam wrote,[67] 'is not that the scientific facts are wrong, but that they don't answer any of the philosophical questions.' Where Vollmer, Lewontin and Thagard and others questioned the facts or the interpretation of the facts used, we now come up against the philosophers who think the facts are right, but that they don't help. 'What is lacking', R. Giere is reported to have said at the beginning of the Ghent Conference on Evolutionary Epistemology,[68] 'is an important problem solved by the clever application of evolutionary epistemology.'

There are a large number of philosophical questions which either have light thrown on them or appear solved when the motley pursuits known as evolutionary epistemology are widened to become philosophical Darwinism. There is no need to go over the ground again. We only have to remind ourselves that philosophical Darwinism has at long last provided a real explanation of the origin of a priori knowledge – where both Plato and Kant had failed and where Locke had been driven to the extraordinarily wrong-headed supposition that it does not exist. It has also at long last provided a very plausible explanation of why the knowledge we have refers to the world we have evolved in. It explains why and how much of epistemology can be naturalised and it explains how knowledge can continue to evolve along Darwinian lines once consciousness and three-dimensional language are in place. It provides a step in the right direction of the problem of reference, of how words hook into things. It does so by showing that 'things' are not known separately and that words do not 'hook' into anything other than what they themselves define and constitute. But, and this is the real contribution, philosophical Darwinism suggests that while the initial condition is defined by the organism-theory, it could not trigger a response or prognosis unless the object or event it defined *really* moved into its ken. Last but not least, evolutionary epistemology underpins Popper's solution of the problem of induction. It shows, as Popper has never tired of arguing, that organisms do not learn by taking instructions and that knowledge is not induced by the world in organisms; but that they learn by making an abundance of proposals, a few of which are selectively retained. Perhaps philosophers are uneasy with these solutions because they do not come, in the main, from philosophers, but from psychologists, ethologists, physicists and biologists like Ghiselin,[69] Donald Campbell,[70] Konrad Lorenz,[71] Karl Popper[72] and Gerhard Vollmer.[73]

Let me examine one particular and very crucial philosophical problem and the answer provided by philosophical Darwinism, the problem of regularities. I begin with Richard Rorty's view that

> there would not have been thought to be a problem about the nature of reason had our race confined itself to pointing out particular states of affairs – warning of cliffs and rain, celebrating individual births and deaths. But poetry speaks of man, birth and death as such, and mathematics prides itself on overlooking individual details. When poetry and mathematics had come to self-consciousness ... the time had come for something general to be said about knowledge and universals.[74]

Rorty believes that the world consists of particular or singular events and that we got unnecessarily confused when we started to think otherwise. Philosophers have argued about this matter ever since Plato, though very few have been driven to take so extreme a view of it as Rorty. Philosophical Darwinism not only shows that Rorty's extreme view is mistaken but also throws an interesting light on a great many of the less extreme views which have been put forward in the course of millennia.

Evolution could not have taken place if there were no regularities in the world. Organisms, even the most primitive organisms, are adapted to features which occur regularly. There is no way in which an organism could store the myriads of particular bits of information necessary for its survival. The organism must store information about regular occurrences. This indicates that regularities really do occur in the world and that Rorty is mistaken in believing that they do not and that we only think they do because we have become confused. Even so, there remains a problem. No two 'regular' occurrences are completely alike. No two flies, even in the frog's eye's perception, are exactly of the same size or shade of black. An organism must therefore be able not to perceive too minutely, lest it really becomes confused. In other words, it must be capable of abstracting the regular features and of disregarding dissimilarities. An organism can only be adapted, no matter how imperfectly, if it has power to abstract.

So what is really in question is not the regularities as such, but the status of the power of abstraction. In Rorty's view, this status ought to be nil because the power to abstract distorts. But if the power to abstract has no status, evolution would not have taken place. Since evolution has taken place, the real question is to find out exactly what status it does have. In the view of many philosophers, the power to abstract is important, but creates a picture of the world which would not exist but for our power to abstract. Its status is therefore considered to be the status of an outsider, of someone who imports an alien and possibly weird faculty into the world. The reason why this weird faculty does not destroy its owner is that the owner ceases – because of this faculty – to be capable of noticing that the world is other than her or his weird faculty suggests. This line of reasoning ought to have been weakened by the sobering consideration that the faculty of abstraction is not peculiar to *homo sapiens* but is present in every single organism, no matter how lowly. But before evolutionary epistemology, not to mention philosophical Darwinism, philosophers have rarely shown interest in the cognitive faculties of non-human organisms. The conclusion we must reach instead is that the faculty of abstraction has the status of a genuine native. Organisms can abstract and disregard similarities because there really are regularities in the world and their adaptation depends on the ability to recognise them and to disregard incidental and accidental features.

The power to abstract could not possibly have been learnt by the 'first' organism by observation, even though it might be conceivable that from then on it could have been passed on by heredity. An organism unable to abstract would have starved to death long before it could have learnt that there really are regularities. In the case of bisexual organisms, there also would have been no offspring at all, because without the power to abstract, no organism could recognise a suitable partner for sexual intercourse. In any case, an organism without power to abstract could not even have learnt from observation, supposing that it might have found somebody to feed it while it was learning. It could not have learnt, because without the power to abstract it could not have known what to learn, that is, what features to disregard as accidental dissimilarities. Hume is famous for believing that such learning is possible. But 'Hume made one uncritical presupposition: that in general certain impressions are given in objective and regular succession',[75] so that the problem of which dissimilarities have to be dismissed would not arise.[76] The only possible answer to the question how the power to abstract could have been acquired is that, since it corresponds to real features of the world, it was acquired in the same way in which any information is acquired. Organisms which by chance mutation had the power survived, while those that did not have it, though they might have been generated or emerged, promptly disappeared again.

The presence in nature of regularities is sufficient explanation of why there are laws of nature which describe causalities and propensities. The distinction between causal laws and ordinary regularities is not absolute, but a matter of degree. If we have two molecules which bond together regularly, we have a regular occurrence. But if we can explain any one such bonding as a particular instance of another regularity, then the first regularity will be 'explained', that is, it can be seen to be due to a causal agency. The employment of the word 'cause' becomes justified once we have a hierarchy of regularities in which the lower regularities can be explained by the higher ones. In this way a statement of a regularity can be converted into a statement of a causal law.

Unfortunately this matter has become very confused by terminological questions. The application of the term 'law' to a regularity goes back to the fifteenth or sixteenth century. It is not absolutely clear why this change in terminology was brought about, but it appears to have been influenced by the theological thought that regularities must have been created or decreed by God and that, therefore, they ought to be called 'laws'.[77] Eventually the theological meaning was forgotten and there emerged instead the notion that the reason why a regularity is called a 'law' is that there is something necessary about it. A regularity may be accidental, like a high correlation between tidal waves on the Californian west coast and the mortality rates of infants in New York. But a causal 'law' had to be necessary in the sense that it could be understood not to be an accidental correlation of events. Whatever the terminological shifts and whatever the reasons for these shifts, there is no great problem as to substance. Regularities can be seen as causal laws as soon as they can be deduced from another regularity. The regularity between tidal waves and infant mortality, no matter how statistically confirmed, is not likely to be so transformed, because there is no known higher regularity from which it could be deduced.

I take it that this view is compatible with David Armstrong's argument that the laws of nature are statements of contingent relations among real universal properties.[78] Armstrong seems to be driven to this rigid distinction between causal laws and mere regularities by the assumption that a causal law is not just a regularity deduced from another regularity; but something more substantial and intelligible than a mere correlation. 'It is a law that Fs are Gs' cannot be analysed[79] to mean that 'all Fs are Gs'. Whether the law-like necessity expressed in the first statement is physical and contingent or logical does not matter all that much. But in order to recognise that it is either the one or the other, one has to recognise the reality of universals. In other words, he insists rightly that to see a statement as a 'law', one has to recognise that there are underlying regularities or 'universals', as he calls them.

By contrast, the views of Nancy Cartwright[80] are incompatible with philosophical Darwinism, which commits us to the proposition that regularities are real and all-pervasive. Nancy Cartwright thinks that it is an unwarranted belief that 'nature is well regulated; in the extreme, that there is a law to cover every case'.[81] She denies that there are any fundamental laws about natural events other than those particular ones which can be established in very special domains like laboratories. These laws, she continues, cannot be held to apply beyond those special domains: 'I imagine that natural objects are much like people in societies. Their behaviour is constrained by some specific laws and by a handful of general principles, but it is not determined in detail, even statistically. What happens on most occasions is dictated by no law at all.' It is a truism to state that we 'know' of laws because of the way we test them, i.e. in laboratories. But the further statement that we have no reason to believe that they obtain outside the laboratory is one of the most stringent forms of nominalist positivism and verificationism I have ever come across. It is impossible to say whether Nancy Cartwright's metaphysical anarchism comes from a failure to see the simple connection between regularities and causal laws and whether she might grant that there are regularities, though no laws, outside laboratories; or whether she is a genuine anarchist. If the former is the case, she is making things too easy for herself by confining her discussion of how the laws of nature lie to a very special and rarefied conception of 'laws', i.e. to one which is divorced from the hierarchical position in a ladder of regularities which we have shown laws to occupy. If the latter is the case, her views are incompatible with evolution and, if she were right, she would not be here with us to argue her case.

VII

Nancy Cartwright's metaphysical anarchism brings us back finally to the problem of realism. At the beginning of this chapter it was argued that if one takes evolution as a cognitive process into account, one is entitled to a certain degree of confidence that our knowledge (provided it is not falsified, that it is formulated as a statement of regularities and that one takes the merely verisimilitudinous quality of it into account) is really realistic; that is, it literally refers to events which are really taking place and is a truly true description of these events. This confidence was based on the realisation that knowers have evolved in the world that is known and are genuine natives, not just tourists. The fact that they have evolved is some proof of the adequacy of their knowledge – which indicates that they may be presumed to know what is really going on.

But this contention is vague, and it is not at all clear what such a statement of 'realism' means. Although philosophers have worried ever since Plato whether the world we perceive or are aware of – not necessarily quite the same things – is 'real' or a figment of our imagination, a dream, or possibly something we have conventionally been conditioned to or have agreed to accept as real, there are no known ways by which these doubts can be removed.

There is a very simple reason for this. All conventional arguments proceed by first looking left to the 'knowledge' and then right to the world and then seek to formulate the relationship in terms of such words as 'refer' or 'describe' or, more colloquially, as 'hook into'. The catch in these procedures is that while one can look to the right – to knowledge – one cannot look to the left unless one has looked to the right first. And if one determines the degree or level of realism of the relationship between right and left by assessing it in terms of what is on the right, one can get nowhere, at least not to a place which would inspire confidence as to the level of realism provided by the right.

Philosophical Darwinism gives some assurance that the regularities we know of really do obtain in nature. It would follow then that the more the regularities and laws of nature we know of cohere, the greater the presumption that they correspond to the world they describe and explain, because it is unlikely that once a certain level of coherence is reached, the overall picture should be *accidentally* true. But the term 'describe' is vague and cannot be literally correct. For how do statements about regularities correspond to the world, and how are we to think of the meaning of 'correspondence'? Certainly not in terms of description; for descriptions are conceptual, and nature is not. There may well be events which we describe as black holes. But they are 'holes' only in a manner of speaking. So whatever the statement 'there are black holes' is, it is not a literal description; and if it does correspond to a reality, it does not do so because it *is* a description. We could think of the correspondence as a mirror. But that does not make much sense either, for mirrors reflect and produce doubles; and the term 'black hole' is not a double of the event it refers to. Right through nature we find that the cognitive relationship represents nature and, in doing so, corresponds but does not mirror it. A hoof represents the steppe over which the horse is running and its contours correspond to the contours of the steppe. But a hoof does not mirror the steppe. Gills represent the oxygen available for breathing under water, but they do not mirror it, at least not literally so. In fact the very expression 'literally true' is a paradox. So all we have initially is that the greater the coherence of the laws of nature, the greater their correspondence and, hence, the closer they are to representing reality. But the word 'represent' somehow begs the question.

Let me start nevertheless with Bas van Fraassen's contention that 'science aims to give us, in its theories, a literally true story of what the world is like; and acceptance of a scientific theory involves the belief that it is true'.[82] Realism then is the view that the events to which regularity statements correspond really take place. In realism, statements about what really takes place are either true or false; and what makes them true or false is something external – that is, not a method of testing them by whether they do or do not produce sense-data or conform to linguistic rules or whatever.[83] A realist believes that science (knowledge) gives us a literally true story of what the world is like. Now my argument is that outside evolutionary epistemology it is impossible to establish realism, even though realism is intuitively quite appealing. If one means by 'realism' that truth-value transcends our ability to determine truth-value, realism cannot be a viable view, for truth-value can have no conceivable meaning over and above our ability to determine it. Nevertheless, realism is commonly so defined. 'Realism is the thesis', writes Colin McGinn,[84] 'that truth or falsity is an epistemically unconstrained property of a sentence. There is nothing in the concept of truth or falsity to exclude the possibility that a sentence may be *unknowably* true or false.' Such and similar reasoning seems patently absurd, and if realism has no better foundation, it is bound to remain questionable. But if we base realism on the consideration that nothing could have evolved had there not been a *real* world to select, realism becomes not only viable but also highly probable.

Some philosophers attempt a generic and reasoned argument for realism. Since realism is said to be vindicated only if it can be established that the truth or falsity of statements is independent of any human way of testing them, i.e. only if they are true or false in virtue of something independent of a human observer, one might expect that, if at all, only generic reasoning can succeed. For such reasoning does not go in search of touchstones of what is and what is not real, but reasons that the events referred to by true knowledge must be real and that it would be some kind of logical contradiction to suppose that they were not. A generic argument in favour of realism is something like a transcendental deduction: one seeks to deduce realism from some condition one thinks one knows. Hilary Putnam, for example, has argued that realism is the only philosophy which does not make the success of science appear a miracle. If we do not suppose that there is a reality to which science corresponds, we cannot account for the success of science. He also put the argument in a slightly different form: 'If we concede that our theories might be wrong, we manifest a realist understanding of truth.'[85]

As against this, Richard Rorty keeps arguing that the argument that science corresponds to reality because it helps us cope is faulty. Science, he maintains, simply helps us cope, full stop. Rorty's argument is not as absurd as it sounds. For one could readily suppose that the reason why science is successful and helps us cope is other than that it corresponds to reality. All one needs is the old conventionalist argument which says that the reason why science works is that there is a convention which disallows any evidence that it does not work. In this way, the success of science can be explained by the simple exclusion of all the cases in which it does not work. As Popper put it:

> For the conventionalist, theoretical natural science is not a picture of nature but merely a logical construction. It is not the properties of the world which determine its construction; on the contrary it is the construction which determines the properties of the artificial world: a world of concepts implicitly defined by the natural laws which we have chosen. It is only *this* world of which science speaks.[86]

Kuhn has used the argument that science (knowledge) helps us cope, and does nothing but, to establish the non-realism of knowledge: 'I can see [in the systems of Aristotle, Newton and Einstein] no coherent direction of ontological development.'[87] On the assumption that knowledge simply helps us cope, and nothing but, Kuhn is undoubtedly right. Kuhn's conventionalism is a form of non-realism and refutes Putnam's argument about miracles. Since the very argument which Putnam thinks leads to realism leads Kuhn to anti-realism, Wittgenstein was perhaps right when he said: 'The most difficult thing in philosophy is to maintain realism without empiricism.'[88] He meant that *generic* arguments are *really* difficult and not likely to succeed. How, indeed, can one hope to establish that there are facts which obtain independently of us and of our capacity to discover them? For realism to be credible, one has to take these facts to be autonomous in that they make no reference to human capacities.

Let me look, therefore, at another type of argument, a non-generic argument in which some kind of (empirical?) proof for realism is looked for.

The second sentence of Quine's *Word and Object* reads as follows: 'Physical things generally, however remote, become known to us only through the effects which they help to induce at our sensory surfaces.' Events which induce these effects are real; and events which do not, are not. These effects are set up as a touchstone of reality. Ayer put the same thought in a slightly different way: 'Our criteria of reality are formed by the way things appear to us. We have nothing else to go by.'[89]

As long as we abide by sensory surfaces, idealists will continue to have a field day, as the long and never-ending procession of anti-realists from Berkeley to Dummett proves. For arguments from sensory surfaces or some other variety of touchstone can and have been used with greater plausibility to demonstrate that scientific statements cannot refer to a real world which exists outside, independently of our perception. Thus Dummett,[90] for example, taking the truth-value of a statement to be the criterion of realism, can show that realism is untenable. Since no statement can have a truth-value independently of our means of knowing that statement, he concludes correctly, that realism is refuted. But Dummett's refutation depends entirely on the supposition that a special touchstone like 'verification' is required to establish realism. Dummett's anti-realism would appear groundless only if one could go back to a different kind of generic argument – an argument which is based on evolution. The search for *touchstones* must inevitably lead to anti-realism and some form or other of idealism and is even more bound to fail than a generic argument. It is not surprising that Quine, when he realised where the touchstone argument was leading to, explicitly disavowed such touchstones as definitions of physical (= real) objects in terms of experience and stated that our trust that science and knowledge refer to events which *really* take place can be no more than a vague and unjustifiable preference:

> As an empiricist I continue to think of the conceptual scheme of science as a tool, ultimately for predicting future experience in the light of past experience. Physical objects are conceptually imported into the situation as convenient intermediaries – not by definition in terms of experience, but simply as irreducible posits comparable, epistemologically, to the gods of Homer. For my part I do, *qua* lay physicist, believe in physical objects and not in Homer's gods; and I consider it a scientific error to believe otherwise. But in point of epistemological footing, physical objects and the gods differ only in degree and not in kind.[91]

In a widely acclaimed book Bas van Fraassen surrenders all claims to realism and argues instead in favour of an anti-realist position. '*Science*', he writes, '*aims to give us theories which are empirically adequate; and acceptance of a theory involves as belief only that it is empirically adequate.*'[92] Van Fraassen calls this 'constructive empiricism' and contrasts it to realism. Science, to be empirically adequate, must 'save the appearances' but does not allow us to suppose that it is literally true. Although the argument is generic, it does depend on the presence of a touchstone. In doing no more than 'saving the appearances', science stands or falls by what can be taken to 'appear'. Van Fraassen calls his view an anti-realist view because we can never know that the theories we have discovered are true. To 'know' so would, of course, involve a prior and independent knowledge of what the world is like, so that we could then compare our theories with what the world is like. He thinks that the idea that science searches for true theories should therefore be replaced by the idea that science searches for theories which are empirically adequate.

Van Fraassen points out[93] that 'science is a biological phenomenon, an activity by one kind of organism which facilitates its interaction with the environment'. It seems that he himself does not take this statement seriously enough. There is a cat in his bag which he is trying hard not to let out. If taken seriously, his argument implies less anti-realism than Fraassen admits to. If scientific theories are designed to be empirically adequate by saving the appearances, then a theory about science should also save the appearances. It so happens that there are appearances called organisms, and *these* appearances can only be saved by the assumption that there must have been a real world which has eliminated and keeps on eliminating less well-adapted organisms and theories in favour of more adapted ones. If there are organisms whose interaction with the environment is facilitated by science, they can only be saved (i.e. any theory about them must be empirically adequate) on the assumption that the reason why they are appearances (i.e. why they are here) is that they have evolved by natural selection. If there had not been a world to eliminate or falsify all those organisms which embodied less truthful theories, these appearances would not be here today. So the rock-bottom in our thinking about realism must be that there have been myriads of 'falsifications' and error eliminations to make our appearance possible. The agent which caused these falsifications and eliminations is what we call the real world.

We could only have evolved in response to that real world. We may be capable of phantasies and dreams. But only a real, independent world could have caused all the eliminations which were necessary for evolution to have taken place and for putting us here to be able to have phantasies and dreams, including phantasies and dreams about a 'real' world. The organisms which have been left over after the real world has done its eliminations must embody theories about the real world which are true, or at least verisimilitudinous. The truth of these embodied theories does not have to wait on our verifications or on certain effects on our sensory surface. For if these theories were not true, the organisms would not be here. There is also some assurance that the regularities disembodied organisms have come to know through the application of the criteria dictated by the minimal ontology necessary for evolution to have taken place really refer to a real world; and, moreover, that they do so because there is something external and not because there is a method of testing them. That world exists independently of our perception, and the truth-value of these theories is independent of our means of knowing it. Its truth-value depends on our evolved presence and our evolved ability to select theories according to the demands of the minimal ontology.

This generic argument is a refutation of Dummett's argument that only statements which are 'independent of our means of knowing' them can 'possess an objective truth-value',[94] for it states that organisms (embodied theories) and theories (disembodied organisms) would not have evolved unless they had some objective truth-value. The generic argument from philosophical Darwinism throws the burden of proof on our existence and removes it from our 'means of knowing' a proof. The argument from evolution favours the thesis that there is a real world to which our knowledge refers and that that world exists the way we think it does, regardless of whether we 'know' of it, i.e. regardless of whether we have or do not have a touchstone. 'It is through the falsification of our suppositions that we actually get in touch with "reality".'[95] The importance of this Popperian argument cannot be exaggerated. Realism is established, it says, because there is an irresistible wall against which all knowledge – whether organisms or theories – beats its head. Such knowledge is not composed, it insists, from particular experiences or verifications of experiences, but consists of what remains when it has beaten its head against the wall. It is made up of all those theories which remain when all sources of criticism have been temporarily exhausted. In this sense, philosophical Darwinism establishes realism. It does so not by looking, like Dummett, for a touchstone, and not by engaging in Putnam's generic reasoning that the successes of science would be a miracle if science did not correspond to a reality. It does so on the ground of the realisation that the only positive evidence we can have for believing that there is a reality out there, and that we must have some knowledge of it, consists in the negative fact that it has brought about all those eliminations without which evolution could not have taken place. Philosophical Darwinism alone satisfies the requirement that there must be something external to us (something other than what we happen to sense or experience or what is induced on our skin) to determine the truth or falsity of statements which refer to that reality. Since our knowledge of evolution, on which these reflections about reality are based, is a theory and a hypothesis to save certain appearances, the realism it leads to should be called 'hypothetical realism'.

NOTES

INTRODUCTION: COGNITIVE CONDITIONS

1 'Logic of Discovery or Psychology of Research?', in I. Lakatos and A. Musgrave (eds), *Criticism and the Growth of Knowledge*, Cambridge, Cambridge University Press, 1970, p. 20; see also G. M. Edelman, *The Remembered Present*, New York, Basic Books, 1989, p. 268: 'Scientific procedure serves to correct perceptual and conceptual errors by communal action and agreement.'

2 *Realism and the Aim of Knowledge*, London, Routledge, 1982, pp. 12–13.

3 See J. Ziman, *Public Knowledge*, Cambridge, Cambridge University Press, 1968.

4 See Peter Munz, 'Taking Darwin Even More Seriously', in K. Hahlweg and C. A. Hooker (eds), *Issues in Evolutionary Epistemology*, New York, SUNY Press, 1989.

5 See, for example, Karl R. Popper, *The Open Society and its Enemies*, Princeton, Princeton University Press, 1950, pp. 416–17.

6 H. R. Maturana, *The Tree of Knowledge*, Boston, Random House, 1987, pp. 45–6.

7 *Knowledge and the Flow of Information*, Oxford, Blackwell, 1981, p. vii.

8 See, e.g., R. Kaspar, 'Die Evolution erkenntnisgewinnender Mechanismen', *Biologie in unserer Zeit*, 1980, 10.

9 *The Remembered Present*, New York, Basic Books, 1989, p. 7.

10 Julian Jaynes, *The Origin of Consciousness in the Breakdown of the Bicameral Mind*, Boston, Houghton Mifflin Co., 1976, speculates that consciousness originated round about 1400 BC. His belief that the anatomical link between the brain's two hemispheres evolved as recently as that century seems bizarre.

11 A. J. Marcel, 'Phenomenal Experience and Functionalism', in A. J. Marcel and E. Bisiach (eds), *Consciousness in Contemporary Science*, Oxford, Clarendon Press, 1988, p. 121. See also L. G. Roberts, 'Machine Perception of Three-Dimensional Solids', in T. Tippett *et al.* (eds), *Optical and Electro-Optical Information Processing*, Cambridge, Mass., MIT Press, 1965.

12 'Scientific Reduction and the Essential Incompleteness of All Science', in F. J. Ayala and T. Dobzhansky (eds), *Studies in the Philosophy of Biology*, London, Macmillan, 1974, p. 273.

13 *Science, Man and Morals*, London, Methuen, 1965, p. 53. The italics are mine and have been added in view of my argument below, that in the first instance, powerful consciousness was more like a by-product and probably an overkill. The point will be that only its retention was no accident.

14 *The Principles of Psychology*, New York, Dover, 1950 reprint, p. 225.

15 *A Theory of Determinism*, Oxford, Clarendon Press, 1988, p. 82.

16 *The Emperor's New Mind*, New York, Oxford University Press, 1989, p. 429. To

understand Penrose correctly, it is necessary to recall that he holds that 'necessary truths' cannot be computed algorithmically.

17 *Animal Thinking*, Cambridge, Mass., Harvard University Press, 1984.
18 'A Modified View of Consciousness', *Psychological Review*, 1969, 76, p. 532. More recently the same view has been expressed by Edelman, *The Remembered Present*, op. cit., passim.
19 S. J. Gould and R. C. Lewontin, 'The Spandrels of San Marco and the Panglossian Paradigm', in E. Sober (ed.), *Conceptual Issues in Evolutionary Biology*, Cambridge, Mass., MIT Press, 1984.
20. 'Comments on Dennett', *Synthese*, 1982, 53, p. 186.
21 J. Levine, 'Review of R. Cummins, "The Nature of Psychological Explanation" ', *Philosophical Review*, 1987, 96, p. 265.
22 *Being-in-the-World*, Cambridge, Mass., MIT Press, 1988.
23 Unfortunately P. M. Churchland, 'Eliminative Materialism and Psychologico-Propositional Attitudes', *Journal of Philosophy*, 1981, 78, pp. 67–90, and *A Neurocomputational Perspective*, Cambridge, Mass., MIT Press, 1990, thinks, on the contrary, that soon neuroscience will make discoveries which will enable us to forget about 'folk-psychology' and propositional attitudes. The description of humans as agents driven by beliefs or desires is folk-psychology which will eventually be shown to be false by its failure to reduce to neuroscience. See also S. Stich, *From Folk-Psychology to Cognitive Science*, Cambridge, Mass., MIT Press, 1983. There is very little in my extensive reading of neuroscience to support such a hope, and much in the opinions of actual neuroscientists to dash it.
24 *Pursuit of Truth*, Cambridge, Mass., Harvard University Press, 1990.
25 *Psychosemantics*, Cambridge, Mass., MIT Press, 1987, p. 79.
26 Ibid.; R. M. Gordon, *The Structure of Emotions*, Cambridge, Cambridge University Press, 1987, commits a real *petitio principii*: taking it for granted that every emotion is sententially unequivocably identifiable, he can show what it is about and what its causes are.
27 The equally inconclusive arguments for and against eliminative materialism are set out in Paul M. Churchland, *Matter and Consciousness*, Cambridge, Mass., MIT Press, 1984, pp. 47–9; and for and against private language in S. Blackburn, *Spreading the Word*, Oxford, Clarendon Press, 1984, Chs 2 and 3. Blackburn's account is biassed against private language and not as even-handed as one is led to expect from the value of his arguments.
28 *The Selfish Gene*, New York, Oxford University Press, 1976, p. 63.
29 *The Remembered Present*, op. cit., p. 92.
30 Oxford, Oxford University Press, 1983; see also M. Levin, 'Why We Believe in Other Minds', *Philosophy and Phenomenological Research*, 1984, 44, p. 347.
31 See, e.g., R. D. Alexander, 'Evolution of the Human Psyche', in P. Mellars and Chris Stringer (eds), *The Human Revolution*, Princeton, Princeton University Press, 1989, pp. 488ff.
32 Edelman, op. cit., pp. 103, 255.
33 *Jenseits von Gut und Böse*, in *Werke*, Berlin, De Gruyter, 1967, Abteilung VI, vol. 2, p. 16.
34 See, e.g., Christine A. Skarda, 'Explaining Behaviour', *Inquiry*, 1986, 29.
35 *Philosophy and the Brain*, Oxford, Oxford University Press, 1987, p. 29.
36 *Rationality*, London, Routledge, 1964, pp. 13–14.
37 *The Selfish Gene*, New York, Oxford University Press, 1976, p. 63.
38 See C. F. Hockett, 'Logical Considerations in the Study of Animal Communication', in W. E. Lanyon and W. B. Tavolga (eds), *Animal Communication*, Washington, DC, Institute of Biological Sciences, 1960.

39 P. Lieberman, *The Biology and Evolution of Language*, Cambridge, Mass., Harvard University Press, 1984, p. 248.
40 See T. P. Waldron, *Principles of Language and Mind*, London, Routledge, 1985, p. 40.
41 Ibid, p. 43.
42 Oxford, Clarendon Press, 1972, pp. 235–9.
43 *Theaitetos* 189e.
44 The preceding sentences are partly copied from and partly adapted from Marcel Proust, *Swann's Way*, Part 1, Engl. transl. and edition of 1966, pp. 58–9, just as on two other occasions, one above and one below, the reader will recognise the use made of two famous lines from *A Midsummer Night's Dream*, V, i.
45 This denial goes as far back as F. W. J. Schelling, who thought that the denial is also proof that language could not have been consciously 'designed'. See also L. L. Whyte, *The Unconscious Before Freud*, London, Tavistock, 1960, p. 126.
46 'Neuropsychology and Consciousness', in C. Blakemore and S. Greenfield (eds), *Mindwaves*, Oxford, Blackwell, 1987, p. 311.
47 'Geological Climates and the Origin of Species', *Quarterly Review*, 1869, 126, p. 391.
48 *The Descent of Man*, London, Murray, 1901, p. 121.
49 See D. Bickerton, *Language and Species*, Chicago, University of Chicago Press, 1990, p. 20.
50 'Evolution of the Human Psyche', in P. Mellars and Chris Stringer (eds), *The Human Revolution*, Princeton, Princeton University Press, 1989, p. 488; see also N. K. Humphrey, 'The Social Function of the Intellect', in P. P. G. Bateson and R. A. Hinde (eds), *Growing Points in Ethology*, Cambridge, 1976.

1 MAN'S GLASSY ESSENCE

1 I follow J. A. Fodor's description of Hume, 'The Mind–Body Problem', *Scientific American*, 1981, 244, No. 1, p. 131, because he is using the up-to-date terms of cognitive psychology.
2 *Intentionality*, Cambridge, Cambridge University Press, 1983, p. 1.
3 See D. R. Griffin, *Animal Thinking*, Cambridge, Mass, Harvard University Press, 1984.
4 Cambridge, Cambridge University Press, 1983, p. 448.
5 E.g., 'A Modified Concept of Consciousness', *Psychological Review*, 1969, 76; 'Toward a Theory of Mind', *Proceedings of the National Academy of Science*, 1969, 63.
6 E.g., 'Toward a Cognitive Theory of Consciousness', in C. Wade Savage (ed), *Perception and Cognition*, Minneapolis, 1978.
7 *Neural Darwinism*, Oxford, Oxford University Press, 1989 and *The Remembered Present*, New York, Basic Books, 1989.
8 *The Remembered Present*, p. 269.
9 Ibid., p. 91.
10 Ibid., p. 93.
11 Ibid., p. 93.
12 Ibid., pp. 56, 243 and *Neural Darwinism*, op. cit., pp. 260, 316.
13 *The Remembered Present*, p. 243.
14 Ibid., p. 56.
15 Op. cit., p. 222.
16 Ibid., p. 271.
17 'A Modified Concept of Consciousness', *Psychological Review*, 1969, 76, p. 536.
18 *Neuronal Man*, Engl. transl., New York, Oxford University Press, 1985, p. 169.

19 Engl. transl., Princeton, Princeton University Press, 1941, p. 68.
20 Ibid., p. 182.
21 Engl. transl., London, Hogarth Press, 1958, pp. 76, 80.
22 Harmondsworth, Penguin, 1959, p. 196.
23 *Philosophical Review*, 1974, 83.
24 *Minds, Brains and Science*, London, BBC, 1984, p. 25.
25 For examples see W. Lyons, *The Disappearance of Introspection*, Cambridge, Mass., MIT Press, 1986, p. 20.
26 'Cerebral Organisation and Behavior', in H. Solomon et al. (eds), *The Brain and Human Behaviour*, Baltimore, Williams & Wilkins, 1956, p. 4.
27 See the experimental results of S. Schachter, *The Psychology of Affiliation*, Stanford, Stanford University Press, 1967, pp. 126–7, which show the wide variety of verbal options available to people physically excited by equal amounts of adrenalin.
28 Peter L. Berger and Th. Luckmann, *The Social Construction of Reality*, Harmondsworth, Penguin, 1971, p. 70.
29 *Philosophical Grammar*, I. 3.
30 The example is not made up but taken from a letter by John R. Searle to the *New York Review of Books*, 1990, XXXVII, No. 10, p. 58, col. 4.
31 D. A. Oakley, 'Animal Awareness', in D. A. Oakley (ed.), *Brain and Mind*, London, Methuen, 1985, p. 148.
32 Changeux, op. cit., pp. 234, 242, 268, 272.
33 K. Lorenz, *On Aggression*, Engl. transl., London, Methuen, 1966, p. 180, and D. R. Griffin, *The Question of Animal Awareness*, New York, Rockefeller Press, 1976, p. 5.
34 *Philosophy and the Brain*, Oxford, Oxford University Press, 1987, p. 74.
35 John Eccles, *The Brain and the Unity of Conscious Experience*, Cambridge, Cambridge University Press, 1965, p. 12.
36 F. de Saussure, *Cours de linguistique générale*, ed. Tullio de Lauro, Paris, Payot, 1972, p. 327.
37 P. M. Churchland and Patricia Smith Churchland, 'Functionalism, Qualia and Intentionality', in J. T. Biro and R. W. Shahan (eds), *Mind, Brain and Function*, Brighton, Harvester, 1982, p. 121.
38 G. Humphrey, *Thinking*, London, Methuen, 1951, Ch. III. See also H. Poincaré, *The Foundations of Science*, Engl. transl., Lancaster, Pa., Science Press, 1913, p. 348.
39 E.g. Daniel Dennett, 'Quining Qualia', in A. J. Marcel and E. Bisiach (eds), *Consciousness in Contemporary Science*, Oxford, Clarendon Press, 1988.
40 This point is well put by Zeno Vendler, *The Matter of Minds*, Oxford, Clarendon Press, 1984, p. 96, when he says that subjective experiences are 'owned'.
41 Howard Gardner, *Frames of Mind*, London, Paladin, 1985, p. 267.
42 *Brainstorms*, Brighton, Harvester Press, 1981, p. 166.
43 *Philosophical Grammar*, ed. Rush Rhees, Oxford, Blackwell, 1974, p. 41.
44 'Toward a Cognitive Theory of Consciousness', in C. Wade (ed.), *Perception and Cognition*, Minneapolis, 1978, p. 222.
45 *Meaning and the Moral Sciences*, London, Routledge, 1978, p. 97.
46 See, e.g., A. J. Marcel, 'Phenomenal Experience and Functionalism', in Marcel and Bisiach (eds), op. cit., p. 128.
47 T. L. S. Sprigge, 'The Importance of Subjectivity', *Inquiry*, 1982, 25, pp. 146–7.
48 *Philosophical Investigations*, I, 246.
49 'In Defence of Eliminative Materialism', in D. Rosenthal (ed), *Materialism and the Mind–Body Problem*, Englewood Cliffs, NJ, Prentice-Hall, 1971, p. 229.
50 Op. cit., 197e.
51 Ibid., 212e.

52 See Peter Munz, 'James Joyce, Myth-Maker at the End of Time', in Donald Ph. Verene (ed.), *Vico and Joyce*, New York: SUNY Press, 1987.
53 J.-K. Huysmans, *Against Nature*, Engl. transl., Harmondsworth, Penguin, 1959, p. 196.
54 *Brainstorms*, p. 171.
55 *Intentionality*, Cambridge, Cambridge University Press, 1983, p. 27.
56 Edelman, *Neural Darwinism*, op. cit., pp. 259–60 and 316.
57 W. H. Harman, 'Alternate Futures and Habitability', in J. Stulman and E. Laszlo (eds), *Emergent Man*, New York, Gordon & Breach, 1973.
58 F. Suppe, *The Structure of Scientific Theories*, Urbana, University of Illinois Press, 1977, p. 3.
59 Cambridge, Mass., MIT Press, 1986, p. 455.
60 E. H. Land, 'The Retinex Theory of Colour Vision', *Scientific American*, 1977, 237, pp. 108–28.
61 Changeux, op. cit., p. 163.
62 Ibid., p. 168.
63 Ibid., p. 109.
64 R. A. Wise, 'The Dopamine Synapse and the Notion of Pleasure Centres in the Brain', *Trends in Neuroscience*, 1980, 3, pp. 91–5.
65 Gardner, op. cit., p. 267.
66 David Marr, 'Early Processing of Visual Information', *Philosophical Transactions of the Royal Society of London*, 1976, Series B, 275, pp. 516–7.
67 Israel Rosenfield, 'Seeing through the Brain', *New York Review of Books*, 1984, 31, No. 15, p. 56.
68 *Philosophical Investigations*, I, 196e.
69 Ibid., 197e.
70 Ibid., 193e.
71 *Mental Models*, Cambridge, Cambridge University Press, 1983, p. 9.
72 'Through a Computer Darkly: Group Selection and Brain Function', *Bulletin of the American Academy of Arts and Sciences*, 1982, 36, no. 1, pp. 21–49. Note the intuitive anticipation of Edelman by A. F. v. Hayek, *The Sensory Order*, London, Routledge 1952, esp. p. 53: 'We do not have first sensations which are then preserved in memory, but it is a result of physiological memory that the physiological impulses are converted into sensations.'
73 See P. Geach, *Mental Acts*, London, Routledge, 1957, section 15.
74 *Studies in Cognitive Growth*, New York, Wiley, 1967, pp. 1–2.
75 Cambridge, Mass., MIT Press, 1983.
76 Ibid., p. 38.
77 *Individuals*, London, Methuen, 1959, p. 110.
78 *Philosophical Investigations*, I, 244.
79 'Studying Human Consciousness Empirically', *Synthese*, 1982, 53, pp. 159–80.
80 A. Hannay, *Mental Images*, London, Allen & Unwin, 1971, p. 99.
81 Quoted in Paul Hoffmann, *Symbolismus*, München, Fink, 1987, p. 122.
82 *Collected Works*, Engl. transl., London, Hogarth Press, 1955, XVII.
83 I. Hacking, *Why Does Language Matter to Philosophy?*, Cambridge, Cambridge University Press, 1975, p. 187.
84 K. R. Popper and J. C. Eccles, *The Self and its Brain*, Berlin, Springer International, 1977.
85 'Self-Consciousness and Intentionality', in G. E. Schwartz and D. Shapiro (eds), *Consciousness and Self-Regulation: Advances in Research*, New York, Plenum Press, 1976, I, p. 84.
86 'Consciousness, Freewill and Personal Identity', in D. A. Oakley and H. C. Plotkin (eds), *Brain, Behaviour and Evolution*, London, Methuen, 1979, p. 223.

87 *Neural Darwinism*, Oxford, Oxford University Press, 1989.
88 *The Emperor's New Mind*, New York, Oxford University Press, 1989.
89 *What Computers Can't Do: A Critique of Artificial Reason*, New York, Harper & Row, 1972.
90 See Patrick H. Wilson's definition of AI as 'the study of ideas that enable computers to be intelligent', *Artificial Intelligence*, 2nd edn, Reading, Mass., Addison-Wesley, 1984, p. 1; and M. L. Minsky's notion that 'artificial intelligence is the science of making machines do things that would require intelligence if done by men'. 'Matter, Mind, and Models', in M. L. Minsky (ed.), *Semantic Information Processing*, Cambridge, Mass., MIT Press, 1968, p. v.
91 J. A. Fodor, 'The Mind–Body Problem', p. 128.
92 H. Putnam, *Mind, Language and Reality*, Cambridge, Cambridge University Press, 1975, Vol. 2, p. xiii.
93 Zenon W. Pylyshyn, 'Minds, Machines and Phenomenology', *Cognition*, 1974–5, 3, p. 61.
94 New York, Thomas Y. Crowell, 1975, p. 198.
95 See J. A. Fodor, 'Methodological Solipsism', *Behavioral and Brain Sciences*, 1980, 3, p. 67.
96 Cambridge, Mass., MIT Press, 1983.
97 To use a phrase which Dennett himself has used to dismiss the Popper–Eccles theory of dualism, *Journal of Philosophy*, 1979, 66, p. 93.

2 THE DUBIOUS CREDENTIALS OF POSITIVISM

1 L. Kolakowski, *Positivist Philosophy*, Penguin, 1972, p. 17.
2 Engl. transl., *Positivism*, Cambridge, Mass., Harvard University Press, 1951.
3 Baltimore, Johns Hopkins Press, 1971.
4 'What Theories are Not', in E. Nagel et al. (eds), *Logic, Methodology and Philosophy of Science*, Stanford, Stanford University Press, 1962.
5 F. Suppe, Introduction to F. Suppe (ed.), *The Structure of Scientific Theories*, Urbana, University of Illinois University Press, 1977, p. 3.
6 Open Court, La Salle, Ill., 1973, p. 5.
7 *An Essay Concerning Human Understanding*, IV, i, 2.
8 Ibid., IV, xxi, 4.
9 Ibid., IV, iv, 3.
10 'Pattern Matching as an Essential in Distal Knowing', in K. R. Hammond (ed.), *The Psychology of Egon Brunswick*, New York, Holt, Rinehart, 1966, p. 85.
11 See Keith Thomas, *Man and the Natural World*, London, Allen Lane, 1983, p. 19.
12 Ibid., p. 39.
13 Ibid., p. 132. For a fuller discussion see A. O. Lovejoy, *The Great Chain of Being*, New York, Harper Torchbooks, 1960, pp. 235ff.
14 *The Newtonian Revolution*, Cambridge, Cambridge University Press, 1980, p. 13.
15 Ibid., p. 99.
16 G. Buchdahl, *The Image of Newton and Locke in the Age of Reason*, London, Sheed & Ward, p. 13.
17 'In the beginning all the world was America', *Two Treatises of Government*, New York, New American Library, p. 343.
18 See Ira O. Wade, *The Intellectual Development of Voltaire*, Princeton, Princeton University Press, 1969, p. 62.
19 London, Dent, Everyman edn, 1914, pp. 64–5.
20 *Oeuvres*, Paris, 1752, III, p. 271.
21 See Katherine George, 'The Civilised West Looks at Primitive Africa: 1400–1800', *Isis*, 1958, 49, pp. 62–72.

22 *Social Science and the Ignoble Savage*, Cambridge, Cambridge University Press, 1976.
23 For the enormous hold the cycles had on the human mind see G. Trompf, *The Idea of Historical Recurrence in Western Thought*, Berkeley, University of California Press, 1979.
24 Meek, op. cit., p. 23 and R. Nisbet, *History of the Idea of Progress*, London, Heinemann, 1969, p. 143.
25 *The Counter-Revolution of Science*, Glencoe, Ill., Free Press, 1952, p. 73.
26 See Peter Munz, *The Shapes of Time*, Middletown, Conn., Wesleyan University Press, 1977.
27 R. W. Southern, *The Shape and Substance of Academic History*, Oxford, Clarendon Press, 1961, p. 11.
28 *Positivism*, op. cit., p. 222.
29 Ibid., p. 223.
30 Loc. cit.
31 *The Poverty of Historicism*, London, Routledge, 1952.
32 Hayek, op. cit., p. 171.
33 *The Positive Philosophy*, Engl. transl., London, Trübner, 1853, p. 200.
34 *August Comte and Positivism*, Boston, Lee & Shepard, 1867, pp. 9–10.
35 Marc Bloch, *Apologie pour l'histoire ou métier de l'historien*, Paris, Armand Colin, 1949, p. xv.
36 Hayek, op. cit., p. 14.
37 See L. Campbell and W. Garnet, *The Life of James Clerk Maxwell*, London, Macmillan, 1884, p. 436.
38 *Scientific Papers*, Cambridge, Cambridge University Press, 1890, II, pp. 361–78. See also p. 328 for Maxwell's blanket rejection of positivism in regard to the relationship between mathematics and minds.
39 See J. Agassi, *Faraday as a Natural Philosopher*, Chicago, University of Chicago Press, 1971, pp. 152ff.
40 *Michael Faraday: A Biography*, London, Chapman, 1965.
41 Ibid., p. 84.
42 Ibid., p. 85.
43 *Über die Erhaltung der Kraft*, Leipzig, Engelman, 1889, pp. 6f.
44 *Vorlesungen über mathematische Physik*, 2nd edn, Leipzig, Teubner, 1877, I, p. 1.
45 *Untersuchungen über die Ausbreitung elektrischer Kraft*, Leipzig, Barth, 1892, p. 23.
46 Ibid., p. 53.
47 See, e.g., E. Cassirer, *The Problem of Knowledge*, New Haven, Yale University Press, 1950, p. 160.
48 See E. Mayr, *The Growth of Biological Thought*, Cambridge, Mass., Harvard University Press, 1973, p. 3.
49 Frances Darwin (ed.), *The Life and Letters of Charles Darwin*, London, Murray, 1887, I, p. 83.
50 D. Hull, *Darwin and his Critics*, Cambridge, Mass., Harvard University Press, 1973, p. 3.
51 Nora Barlow (ed.), *Charles Darwin, The Autobiography*, London, Collins, 1958, p. 120.
52 See John Blackmore, *Ernst Mach*, Berkeley, University of California Press, 1972, p. 120 and Cassirer, op. cit., pp. 84–5, 89.
53 Blackmore, op. cit., pp. 214–6.
54 W. Heisenberg, *Der Teil und das Ganze*, München, Piper, 1969, p. 92. For comment see B. Kanitscheider, *Das Weltbild Albert Einsteins*, München, Beck, 1988, p. 83 and Mario Bunge, *Foundations of Physics*, Berlin, Springer, 1967.
55 G. Radnitzky, *Contemporary Schools of Metascience*, 3rd edn, Chicago, H. Regnery, 1973, p. 81.

56 Ibid., p. 69.
57 Peter Munz, 'Finches, Fossils and Foscarini', *New Zealand Journal of History*, 1980, 14, pp. 132–52.
58 S.J. Gould, *Time's Arrow, Time's Cycle*, Cambridge, Mass., Harvard University Press, 1987, p.105, has drawn attention to the ambiguity of the term 'uniformitarianism' in Lyell, who used it now to indicate gradual, uniform development and now to indicate that the same general laws of physics apply in all geological ages.
59 v. Mises, op. cit., p. 81.
60 Ibid., p. 86; the first stage was Hume.
61 E. Mach, 'Introduction' to J. B. Stallo, *Die Begriffe und Theorien der modernen Physik*, German transl., Leipzig, Barth, 1911.
62 Ibid. p. xii.
63 G. M. Edelman, *Neural Darwinism*, Oxford, Oxford University Press, 1989, p. 316.
64 Quoted in Kolakowski, *Positivist Philosophy*, op. cit., p. 146.

3 THE LURE OF SOCIOLOGY

1 *Dawn*, sect. 516.
2 *The Varieties of Reference*, Oxford, Clarendon Press, 1982.
3 *A Theory of Semiotics*, Engl. transl., Bloomington, Indiana University Press, 1977.
4 The argument is parodied in Tom Stoppard's *Jumpers*. The philosopher, half Wittgenstein, half Ayer, whose wife, raped by a detective, is crying 'Help!', is consoled by a psychiatrist: 'Don't worry! In the profession we interpret this as a cry for help!'
5 Stanley Fish, quoted by F. Crews, *Skeptical Engagements*, New York, Oxford University Press, 1986, p. 125.
6 F. W. Hegel, *System der Philosophie*, Part III, 'Die Philosophie des Geistes', ed. H. Glockner, Jubiläumsausgabe, X, pp. 282–3. See also *Phenomenology of the Spirit*, sects 187–9. Incomprehensibly optimistic, Hegel actually believed that the transformation of the struggle into rule following would take place as soon as there were more than two people present.
7 All three reviews are reprinted in G. Pitcher (ed.), *Wittgenstein: The Philosophical Investigations*, London, Macmillan, 1968.
8 'The New Idealism - Cause and Meaning in the Social Sciences', in I. Lakatos and A. Musgrave (eds), *Problems in the Philosophy of Science*, Amsterdam, North Holland Publishing Co., 1965.
9 Oxford, Clarendon Press, 1968.
10 Oxford, Blackwell, 1982.
11 *Proceedings of the 2nd International Symposium of Kirchberg*, Vienna, Holder, Pichler, Tempsky, 1978, p. 74.
12 London, Macmillan Press, 1983.
13 Barry Barnes, 'On the Extensions of Concepts and the Growth of Knowledge', *Sociological Review*, 30, 1982, p. 35.
14 *The Social Construction of Reality*, Harmondsworth, Penguin, 1971, p. 70.
15 W. v. O. Quine, *Ontological Relativity*, New York, Columbia University Press, pp. 86–7.
16 *The Enlightenment*, New York, Knopf, 1967, I, p. 141.
17 *The History of England*, London, T. Cadell, 1789, Vol. 8, p. 334. See p. 87 above.
18 '... und sehe, dass wir nichts wissen können, das will mir schier das Herz verbrennen ... '.
19 See R. L. Trivers, 'The Evolution of Reciprocal Altruism', *Quarterly Review of Biology*, 1971, 46, pp. 35–58 and Peter Munz, 'Taking Darwin Even More

Seriously', in K. Hahlweg and C. A. Hooker (eds), *Issues in Evolutionary Epistemology*, New York, SUNY Press, 1989, p. 286.

20 The word 'historism' should not be confused with the term 'historicism'. The first term refers to the method of closed circles and the avoidance of reference to anything outside those circles. It is very similar to the post-modernist denial that there are meta-narratives. The second term denotes the opinion that there are developmental laws which govern the process of history and in particular, the succession of these closed circles. Historicism is diametrically opposed to historism, according to which systems do not stand in an intelligible relationship to one another. The two words have nothing in common except the first seven letters.

21 *Ritual, History and Power*, London, Athlone Press, 1989, p. 2.

22 Where Malinowski's functionalism had been used strictly to explain synchronic relationships, C. Renfrew, *The Emergence of Civilisation in the Cyclades*, London, Methuen, 1972 has shown that such functionalism can also be used fruitfully to explain the emergence of cultural systems, that is, diachronically.

23 See Peter Munz, *Our Knowledge of the Growth of Knowledge*, London, Routledge, 1985, pp. 132–45.

24 E. Gellner, *Cause and Meaning in the Social Sciences*, London, Routledge, 1973, Ch. 2, notes 3 and 4; S. Lukes, *Emile Durkheim*, London, Allen Lane, 1973, p. 437.

25 *Tractatus*, 4.1122.

26 Colin McGinn, *Wittgenstein on Meaning*, Oxford, Blackwell, 1984, pp. 49–50, argues, not very convincingly, that there is no connection between Wittgenstein's view that when we are speaking we are following rules and Wittgenstein's view that there is no private language in which we can express immediate, private subjectivity. McGinn believes that it is wrong to read Wittgenstein as saying that we follow rules because we cannot use a private language. If McGinn is right, Wittgenstein's thought would be trivial and hardly worth the attention it has received! Theories of rule following and theories which deny that there is a private language as direct expression of inner subjectivity have been separately around for a long time. What makes Wittgenstein interesting is that he connected the two.

27 In this form it is not to be found in Wittgenstein, but in Kuhn and, especially, in Feyerabend.

28 *The False Prison*, Oxford, Clarendon Press, 1988, p. 458.

29 Ibid., pp. 509–10. See also S. Kripke, *Wittgenstein on Rules and Private Language*, Oxford, Blackwell, 1982, for a slight scepticism about such a reading of Wittgenstein's meaning.

30 I am never quite sure whether this conclusion amounts to saying that those sentences must also be true in that community, or not. Merrill B. and Jaako Hintikka, *Investigating Wittgenstein*, Oxford, Blackwell, 1986, p. 216f., have argued very unconvincingly that for Wittgenstein the 'main function of language-games' is to establish vertical connections between language and reality. They maintain that it was Wittgenstein's view that speaking a language is only part of the language-game, the other part being the establishment of reference to reality. They misquote Wittgenstein, who wrote explicitly in *Philosophical Investigations*, I, 23, that a language-game is part of a form of life; and not, as the Hintikkas maintain, part of speaking a language.

31 Quoted by L. Kolakowski, *Religion*, London, Fontana, 1982, p. 61. For similar remarks by Wittgenstein, see *Zettel*, par. 134.

32 *De Dignitate et Augmentis Scientiarum*, IV, 433.

33 Quoted by M. Levin, 'Caring New World', *The American Scholar*, Winter 1988, p. 101.

34 'Cognitive Styles in Geology', in Mary Douglas (ed.), *Essays in the Sociology of Perception*, London, Routledge, 1982.
35 F. Manuel, *A Portrait of Isaac Newton*, Cambridge, Mass., Belknap Press, 1968, p. 84.
36 S. J. Gould, *Wonderful Life*, London, Hutchinson Radius, 1989.
37 *Progress and its Problems*, Berkeley, University of California Press, 1977, p. 202.
38 See, e.g., C. Behan McCullagh, 'The Intelligibility of Cognitive Relativism', *The Monist*, 1984, 67, p. 327. See also B. Barnes and D. Bloor, 'Relativism, Rationalism and Sociology of Knowledge', in M. Hollis and S. Lukes (eds), *Rationality and Relativism*, Oxford, Blackwell, 1982, p. 27: 'For the relativist, there is no sense attached to the idea that some standards or beliefs are really more rational as distinct from merely accepted as such.'
39 D.Bloor, *Knowledge and Social Imagery*, London, Routledge, 1976.
40 *Morality*, Harmondsworth, Penguin, 1973, p. 37. Donald Davidson, 'On the Very Idea of a Conceptual Scheme', *Proceedings and Addresses of the American Philosophical Association*, XLVII, 1974. See also D. Sperber, 'Apparently Irrational Beliefs', in M. Hollis and S. Lukes (eds), *Rationality and Relativism*, Oxford, Blackwell, 1982, p. 159, and S. Lukes, 'Relativism in its Place', in Hollis and Lukes (eds), op. cit., p. 262.
41 See K. R. Popper, *The Poverty of Historicism*, London, Routledge, 1957.
42 R. G. Collingwood showed great interest in Spengler: 'Oswald Spengler and the Theory of Historical Cycles', *Antiquity*, 1927, I. Collingwood's own view of the history of knowledge had led him to the conclusion that all knowledge consisted of answers to questions and that those questions changed, incomprehensibly and unrelatedly, from time to time.
43 *Philosophical Investigations*, Part II, xi, p. 224. Though the remark may be cryptic, there is no further explanation of it in Garth Hallett, *A Companion to Wittgenstein's Philosophical Investigations*, Ithaca, Cornell University Press, 1977, p. 731. Henry le Roy Finch, *Wittgenstein – The Later Philosophy*, Atlantic Highlands, Humanities Press, 1977, p. 76, is uncritically lyrical about Wittgenstein's concept of spontaneity.
44 *The Structure of Scientific Revolutions*, 2nd edn, Chicago, Chicago University Press, 1970, p. 151.
45 See J. Piaget, *Structuralism*, London, Routledge, 1968, p. 129.

4 THE NATURE OF THE MIRROR

1 *An Essay Concerning Human Understanding*, I, ii, 1.
2 See E. Cassirer, *Kant's Life and Thought*, Eng. transl., New Haven, Yale University Press, 1981, p. 186.
3 Harlow, Longman, 1986.
4 See examples listed by S. Körner in Cassirer, op. cit., pp. x–xi.
5 *History of Western Philosophy*, London, Allen & Unwin, 1946, p. 740.
6 *Werke*, Frankfurt a.M., Insel, 1958, III, pp. 337–8.
7 See A. Gulyga, *Kant*, Engl. transl., Boston, Birkhäuser, 1985, pp.16ff., and Cassirer, op. cit., p. 356, for Kant's ideas of evolution and their limits.
8 *Notebook*, M 128, in H. E. Gruber, *Darwin on Man*, New York, Dutton, 1974, p. 290.
9 Ibid., M 84, p. 281.
10 E. Mayr, *Toward a New Philosophy of Biology*, Cambridge, Mass., Belknap Press, 1988, p. v.
11 4.1222. See also C. Burret (ed.), *L. W. Wittgenstein: Lectures and Conversations*,

Berkeley, University of California Press, 1967, p. 26. It appears that Wittgenstein remained altogether sceptical of evolution because it was, he said, impossible actually to *watch* how speciation occurred. Perhaps it was this same scepticism which predisposed him to believe that language-games did not evolve but emerged 'spontaneously'. See *Philosophical Investigations*, Engl. transl., New York, Macmillan, 1953, Part II, xi, p. 224e and P. Munz, 'The Fly in the Bottle', *Philosophy of the Social Sciences*, 1987, 17, p. 72. Darwin, of course, did make an impression on both Spencer and Bergson; but this impression led into directions which are not relevant to our present concerns. See P. J. Bowler, *Evolution: The History of an Idea*, Berkeley, University of California Press, rev. edn, 1989, pp. 237ff.

12 'Kants Lehre vom Apriorischen im Lichte gegenwärtiger Biologie', *Blätter für deutsche Philosophie*, 1941, 15, 94–125.

13 See, e.g., R. Norman, *Hegel's Phenomenology*, London, Sussex University Press, 1976, p. 11.

14 Popper acknowledges that his approach is independent. Though the term 'evolutionary epistemology' is due to Donald T. Campbell, the idea itself goes back to the end of the nineteenth century. See K. R. Popper, *Objective Knowledge*, Oxford, Clarendon Press, 1972, p. 67.

15 Popper, op. cit., p. 149.

16 *The Poverty of Historicism*, London, Routledge, 1957, p. 106.

17 *Unended Quest*, London, Fontana, 1976, sect. 37.

18 Oxford, Clarendon Press, 1972, p. 241.

19 See, e.g., Michael Ruse's running battle against Popper, *Is Science Sexist?*, Dordrecht, Reidel, 1981, Ch. 3, pp. 65–84, and the frequent asides by R. C. Lewontin, as in 'Fallen Angels', *New York Review of Books*, 1990, XXXVII, No. 10, p. 3, col. 1 or the dismissive comment by Ernst Mayr, *Toward a New Philosophy of Biology*, Cambridge, Mass., Belknap Press, 1988, p. v.

20 'Antibodies and Learning: Selection versus Instruction', in Gardner C. Quarton *et al.* (eds), *The Neurosciences*, New York, The Rockefeller University Press, 1967, p. 201.

21 G. M. Edelman, *Neural Darwinism*, Oxford, Oxford University Press, 1989, p. 317.

22 See J.-P. Changeux, *Neuronal Man*, Engl. transl., New York: Oxford University Press, 1986, pp. 280f. and M. Piatelli-Palmerini, 'The Rise of Selective Theories', in W. Demopoulos and A. Marras (eds), *Language Learning and Concept Acquisition*, Norwood, NJ, Ablex Pub. Co., 1986.

23 See, e.g., E. S. Reid, 'Why Ideas are not in the Mind', in Abner Shimony and Debra Nails (eds), *Naturalistic Epistemology*, Dordrecht, Reidel, 1987, p. 217.

24 See, e.g., B. M. Hessen, *The Social and Economic Roots of Newton's 'Principia'*, Engl. transl., New York, Fertig, 1971; G. Freudenthal, *Atom and Individual in the Age of Newton*, Dordrecht, Reidel, 1986; Margaret Jacob, *The Newtonians and the English Revolution*, Ithaca, Cornell University Press, 1976; and last, but not least, R. K. Merton, 'Science, Technology and Society in 17th Century England', *Osiris*, 1938, 4.

25 See Alvin I. Goldman, *Epistemology and Cognition*, Cambridge, Mass., Harvard University Press, 1986, p. 152.

26 In P. A. Schilpp (ed.), *The Philosophy of Karl Popper*, La Salle, Ill., Open Court, 1974.

27 Stuttgart, Hirzel.

28 Hamburg, Parey.

29 R. J. Riedl and F. Kreuzer (eds), *Evolution und Menschenbild*, Hamburg, Hoffmann und Campe, 1983; K. Lorenz and Franz M. Wuketits (eds), *Concepts and Approaches in Evolutionary Epistemology*, Dordrecht, Kluwer, 1984; K. Hahlweg and C. A.

Hooker (eds), *Issues in Evolutionary Epistemology*, Albany, SUNY Press, 1989; H. C. Plotkin (ed.), *Learning, Development and Culture: Essays in Evolutionary Epistemology*, Chichester, Wiley, 1982; R. Riedl and F. M. Wuketits (eds), *Die Evolutionäre Erkenntnistheorie*, Hamburg, Parey, 1987; W. Callebaut and R. Pinxton (eds), *Evolutionary Epistemology*, Dordrecht, Reidel, 1987; N. Rescher (ed.), *Evolution, Cognition, and Realism: Studies in Evolutionary Epistemology*, Lanham, Md., University Press of America, 1990.

30 The field is briefly and inconclusively surveyed by Michael Ruse, *Philosophy of Biology Today*, New York, SUNY Press, 1988, Ch. 8.

31 See Donald T. Campbell, 'Epistemological Roles for Selection Theory', in Rescher (ed.), op. cit.

32 See J. T. Bonner, *The Evolution of Complexity by Means of Natural Selection*, Princeton, Princeton University Press, 1988.

33 Changeux, op. cit., p. 272; see also p. 249.

34 'Through a Computer Darkly', *Bull. of the American Academy of Arts and Sciences*, 1982, 36, no. 1, pp. 21–49.

35 *Umwelt und Innenwelt der Tiere*, Berlin, Springer, 1920.

36 *Neural Darwinism*, p. 245.

37 Lorenz, *Behind the Mirror*, Engl. transl., London, Methuen, 1977, p. 54.

38 R. Riedl, 'Über die Biologie des Ursachen-Denkens', *Mannheimer Forum*, 1978/79, p. 22.

39 See N. Bischof, *Das Rätsel des Oedipus*, München, Piper, 1985, p. 154.

40 Theories of reference are developed by the aristocrats of the realm of philosophy. See, e.g., Gareth Evans, *The Varieties of Reference*, Oxford, Clarendon Press, 1982 and Saul Kripke, *Naming and Necessity*, Oxford, Blackwell, 1980. Quite recently, Stephen Schiffer, one of the leading members of this aristocracy, has given up in despair: *Remnants of Meaning*, Oxford, Clarendon Press, 1988.

41 *Pursuit of Truth*, Cambridge, Mass., Harvard University Press, 1990, p. 1.

42 *The Tree of Knowledge*, Boston, Mass., Random House, 1987.

43 W. R. Ashby, *An Introduction to Cybernetics*, New York, Wiley & Sons, 1965, pp. 1, 4, 60.

44 R. Gregory, *Eye and Brain*, New York, McGraw-Hill, 1973, p. 94.

45 *From a Logical Point of View*, Cambridge, Mass., Harvard University Press, 1971, p. 43.

46 *Objective Knowledge*, Oxford, Clarendon Press, 1972, Ch. 3.

47 Popper has clarified the precise differences between Hegel's objective spirit and his own reified theories, ibid., pp. 125–6.

48 'Perception from an Evolutionary Point of View', *Journal of Philosophy*, 1971, 68, p. 574.

49 D. Stenhouse, *The Evolution of Intelligence*, London, Allen & Unwin, 1973, p. 67.

50 The view that consciousness is the helmsman or charioteer is as old as Plato and lasted until the Romantics. But the Romantics, instead of sizing consciousness up correctly, simply stood Plato on his head and considered consciousness to be the enemy of life.

51 François Jacob, 'Evolution and *bricolage*', *Science*, 1977, 196.

52 Karl Popper, 'Replies to My Critics', in P. A. Schilpp (ed.), *The Philosophy of Karl Popper*, La Salle, Ill., Open Court, 1974, p. 1112. Popper tends to focus on the mere ability of language to negate. Insects which use mimicry and antelopes which deceive pursuers by darting to the left when they are heading to the right can also 'negate'. Three-dimensional language does more because it enables its user to invent a positive 'alternity' rather than merely negate a reality. See G. Steiner, *After Babel*, Oxford, Oxford University Press, 1975, p. 223.

53 *Objective Knowledge*, p. 261.

54 See P. Munz, 'Taking Darwin Even More Seriously', in K. Hahlweg and C.A. Hooker (eds), *Issues in Evolutionary Epistemology*, op. cit.
55 'Organism and Environment', in Plotkin (ed.), *Learning, Development and Culture*, p. 164.
56 See R. Dawkins, *The Blind Watchmaker*, Harlow, Longman, 1986, p. 258.
57 *Progress and its Problems*, Berkeley, University of California Press, 1977, p. 6.
58 See E. Sober, 'Revisability, A Priori Truth and Evolution', *Australasian Journal of Philosophy*, 1981, 59, p. 84.
59 *The Structure of Scientific Revolutions*, 2nd edn., Chicago, University of Chicago Press, 1970, p. 77.
60 E.g., most recently in *Pursuit of Truth*, Cambridge, Mass., Harvard University Press, 1990.
61 J. D. Barrow, *The World Within the World*, Oxford, Oxford University Press, 1990, pp. 345–52, discusses the importance of aesthetic criteria.
62 *Logik der Forschung*, Vienna, Springer, 1935.
63 Oxford, Clarendon Press, 1986, pp. 68ff.
64 D. C. Stove, *The Rationality of Induction*, Oxford, Clarendon Press, 1986.
65 *Meaning and the Moral Sciences*, London, Routledge, 1978, p. 9.
66 'Confutation of Convergent Realism', in J. Leplin (ed.), *Scientific Realism*, Berkeley, University of California Press, 1984, p. 232.
67 *Conjectures and Refutations*, London, Routledge, 1963, pp. 232–3 and *Objective Knowledge*, Oxford, Clarendon Press, 1972, pp. 48 ff.
68 *Objective Knowledge*, p. 57.
69 *Conjectures and Refutations*, p. 232.
70 *Objective Knowledge*, p. 48. See also *Conjectures and Refutations*, p. 233.
71 For examples see *British Journal for the Philosophy of Science*, 1974, 25, pp. 155–88.
72 *Conjectures and Refutations*, p. 232.
73 'Neurological Embodiments of Belief and the Gaps in the Fit of Phenomena to Noumena', in Abner Shimony and Debra Nails (eds), *Naturalistic Epistemology*, Dordrecht, Reidel, 1987.

5 THE VIEW FROM SOMEWHERE

1 See K. R. Popper, *A World of Propensities*, Bristol, Thoemmes, 1990.
2 The expression is borrowed from Thomas Nagel, *The View from Nowhere*, New York, Oxford University Press, 1986.
3 'Anti-Spengler' (1921), Engl. transl. in M. Neurath and R. S. Cohen (eds), *Empiricism and Sociology*, Dordrecht, D. Reidel, 1973, p. 192.
4 'A Free Man's Worship', in B. Russell, *Mysticism and Logic*, London, Longmans, 1918, pp. 47–8. Over half a century later, these words were echoed by Charles Monod: 'Man must realise that he lives, like a gipsy, on the boundary of an alien world; a world that is deaf to his music and as indifferent to his hopes as it is to his sufferings and his crimes.' *Chance and Necessity*, Engl. transl., New York, Knopf, 1972, p. 177.
5 *The Diary of Antoine Roquentin*, Engl. transl., London, John Lehmann, 1949, p. 176.
6 *The Remembered Present*, New York, Basic Books, 1989, p. 254.
7 See G. Vollmer, *Was können wir wissen?*, Stuttgart, Hirzel, 2 vols., 1986, 2, p. 61.
8 See Ilya Prigogine, 'Die physikalisch-chemischen Wurzeln des Lebens', in H. Meier (ed.), *Die Herausforderung der Evolutionsbiologie*, München, Piper, 1988.
9 Cambridge, Cambridge University Press, 1985, Ch. 2.
10 Ibid., p. 226.

11 See Ernst Mayr, *Towards a New Philosophy of Biology*, Cambridge, Mass., Belknap Press, 1988, pp. 8–23. However, R. C. Lewontin, 'The Organism as the Subject and Object of Evolution', *Scientia*, 1983, 118, pp. 65–6, contrasts the transformational theories of change in physics and chemistry with the principle of variational change which obtains in biology.
12 Op. cit., p. 226.
13 Apart from numerous technical papers and books, see especially *Order out of Chaos: Man's New Dialogue with Nature*, Engl. transl., Boulder, Shambalah, 1984.
14 Apart from numerous technical papers, see *Laws of the Game: How the Principles of Nature Govern Chance*, Engl. transl., New York, Knopf, 1981, Ch. 8, especially section 4.
15 *What is Life?*, Cambridge, Cambridge University Press, 1945; and as Canto paperback, Cambridge, 1992.
16 Prigogine, op. cit., pp. 143, 190–1; see also his 'Time, Irreversibility and Structure', in J. Mehra (ed.), *The Physicist's Conception of Nature*, Dordrecht, Reidel, 1973, p. 561.
17 Ilya Prigogine, Gregoire Nicolis and Agnes Babloyantz, 'Thermodynamics of Evolution', *Physics Today*, December 1972, pp. 43–4.
18 *From Being to Becoming*, San Francisco, W. H. Freeman, 1980, p. 78.
19 *Physics Today*, December 1972, p. 41. See also B.-O. Küppers, *Information and the Origin of Life*, Engl. transl., Cambridge, Mass., MIT Press, 1990.
20 J. S. Wicken, 'Thermodynamics, Evolution and Emergence', in B. Weber *et al.* (eds), *Entropy, Information and Evolution*, Cambridge Mass., MIT Press, 1988, p. 153.
21 The main work of this school of thought is D. R. Brooks and E. O. Wiley, *Evolution as Entropy*, Chicago, University of Chicago Press, 1986.
22 It is difficult to understand how Michael Ruse, *Philosophy of Biology Today*, Albany, SUNY Press, 1988, p. 39, can say that in turning away from selection processes, these biologists are following the non-equilibrium ideas of Prigogine.
23 Mayr, op.cit., p. 163.
24 E. O. Wiley and D. R. Brooks, 'Victims of History – A Nonequilibrium Approach to Evolution', *Systematic Zoology*, 1982, 31, p. 17.
25 *The View from Nowhere*, p. 78.
26 Cp. B. Kanitscheider, *Kosmologie*, Stuttgart, Reclam, 1984, pp. 273–5.
27 *Ontological Relativity and Other Essays*, New York, Columbia University Press, 1969, p. 82.
28 Ibid., pp. 82–3.
29 See, e.g., Christine A. Skarda, 'Explaining Behavior: Bringing the Brain Back In', *Inquiry*, 1986, 29, pp. 187–202.
30 See P. Munz, *When the Golden Bough Breaks*, London, Routledge, 1973, Chs 7–8.
31 Charles Darwin, *Notebooks*, M 84 or M 128, in H. E. Gruber, *Darwin on Man*, New York, Dutton, 1974, before the remark on Locke. P. H. Barrett, 'Darwin's Early and Unpublished Notebooks', in H. E. Gruber (ed.), *Darwin on Man*, New York, Dutton, 1974, p. 281.
32 *The Structure of Biological Science*, Cambridge, Cambridge University Press, 1985, p. 1.
33 Ibid., p. 11.
34 See Peter Munz, 'Philosophy and the Mirror of Rorty', *Journal of the Philosophy of the Social Sciences*, 1984, 14, pp. 195–238, also as Ch. XVI in G. Radnitzky and W. W. Bartley (eds), *Evolutionary Epistemology, Theory of Rationality and the Sociology of Knowledge*, La Salle, Ill., Open Court, 1987.
35 *Biologie der Erkenntnis*, Berlin, Parey, 1979.
36 *Evolutionäre Erkenntnistheorie*, Stuttgart, Hirzel, 1975.

37 *Human Understanding*, Oxford, Clarendon Press, 1972.
38 *Science as a Process*, Chicago, University of Chicago Press, 1988.
39 See, e.g., L. Kolakowski, *Husserl and the Search for Certitude*, New Haven, Yale University Press, 1975, p. 15; G. Vollmer, *Evolutionäre Erkenntnistheorie*, 3rd edn, Stuttgart, Hirzel, 1981, p. 177.
40 *Behind the Mirror*, Engl. transl., London, Methuen, 1977.
41 Op. cit., vol. 2, pp. 39–52.
42 'What Evolutionary Epistemology is Not', in W. Callebaut and R. Pinxton (eds), *Evolutionary Epistemology*, Dordrecht, Reidel, 1987.
43 *Mind from Matter?*, Oxford, Blackwell, 1986, p. 245.
44 See Peter Munz, *Our Knowledge of the Growth of Knowledge*, London, Routledge, 1985, p. 65 and the literature cited in note 51 there.
45 *Evolutionäre Erkenntnistheorie*, 3rd edn, Stuttgart, Hirzel, 1981, p. 78.
46 Leon N. Cooper, *An Introduction to the Meaning and Structure of Physics*, New York, Harper & Row, 1968, p. 510.
47 Paul Thagard, 'Against Evolutionary Epistemology', *Philosophy of Science Association*, 1980, 1, pp. 188–9; R. C. Lewontin, 'Organism and Environment', in H. C. Plotkin (ed.), *Learning, Development and Culture*, New York, Wiley & Sons, 1982, p. 154. Thagard's critique, far from receiving critical discussion, has been thoughtlessly described as a 'hatchet job' by Ch. Hookway, 'Naturalism, Fallibilism and Evolutionary Epistemology', in Ch. Hookway (ed.), *Minds, Machines and Evolution*, Cambridge, Cambridge University Press, 1984, p. 14.
48 A. Kantorovich and Y. Ne'eman, 'Serendipity as a Source of Evolutionary Progress in Science', *Studies in History and Philosophy of Science*, 20, 1989, p. 508. See also Donald T. Campbell, 'Unjustified Variation and Selective Retention in Scientific Discovery', in F. J. Ayala and T. Dobzhansky (eds), *Studies in the Philosophy of Biology*, New York, Macmillan, 1974.
49 'Has the Theory of Evolution any Relevance to Philosophy?', *Ratio*, 1987, 29, p. 28.
50 Thagard, op. cit., p. 190; Lewontin, op. cit., p. 164.
51 Thagard, op. cit., p. 191.
52 Lewontin, op. cit., pp. 157–9.
53 *The Logic of Scientific Discovery*, Engl. transl., London, Hutchinson, 1959, p. 42.
54 Ibid., p. 108.
55 *The Structure of Scientific Revolutions*, 2nd edn, Chicago, Chicago University Press, 1970, p. 77.
56 Ibid., p. 1.
57 See Peter Munz, *The Shapes of Time*, Middletown, Conn., Wesleyan University Press, 1977, pp. 204f. and Munz, *Our Knowledge of the Growth of Knowledge*, pp. 108–10. It is surprising that Larry Laudan, *Progress and its Problems*, Berkeley, University of California Press, 1977, p. 158, believes that one can distinguish between 'the actual past of science' and 'the writings of historians about the past'.
58 *Towards an Historiography of Science*, S-Gravenhage, Mouton & Co., 1963.
59 *Scrutinizing Science*, Dordrecht, Kluwer, 1988, p. 5.
60 In a personal letter to me of 10 November 1976, Kuhn disarmingly admitted that he had allowed a 'bite' to history which he had denied to science.
61 Op. cit., p. 151. With this insight, Kuhn reaffirmed, though in a more humane manner, the epistemological wisdom of Louis IX, the saintly king of France (1226–70) who had recommended that one actually kill people who were living under a different paradigm. Jean Sire de Joinville, *The History of St Louis*, Engl. transl., London, Oxford University Press, 1983, Ch. 16.
62 'Theory Change and the Indeterminacy of Reference', *Journal of Philosophy*, 1973, 70, pp. 462–81.

63 Oxford, Clarendon Press.
64 Chicago, University of Chicago Press.
65 Oxford, Clarendon Press.
66 Ibid., pp. 242–4.
67 'Why Reason Can't be Naturalised', in K. Baynes *et al.* (eds), *Philosophy: End or Transformation*, Cambridge, Mass., MIT Press, 1987, p. 226.
68 Callebaut and Pinxton (eds), *Evolutionary Epistemology*, p. 3.
69 Michael T. Ghiselin, *The Triumph of the Darwinian Method*, Berkeley, University of California Press, 1969.
70 Donald T. Campbell, 'Evolutionary Epistemology', in P. A. Schilpp (ed.), *The Philosophy of Karl Popper*, La Salle, Ill., Open Court, 1974.
71 *Behind the Mirror.*
72 *Objective Knowledge.*
73 Vollmer, *Evolutionäre Erkenntnistheorie*, 3rd edn, and *Was können wir wissen?*.
74 *Philosophy and the Mirror of Nature*, Princeton, Princeton University Press, 1979, p. 38.
75 Ernst Cassirer, *Kant's Life and Thought*, Engl. transl., New Haven, Yale University Press, 1981, p. 186.
76 It has since been argued by Galen Strawson, *The Secret Connection: Causation, Realism and David Hume*, Oxford, Clarendon Press, 1988, that Hume had no such view, and that he is wrongly famous for it. But since it is the view rather than Hume which is interesting, one must suggest, if she is right, that we now cease to read Hume and start studying the person who first put forward this view.
77 See E. Zilsel, 'The Genesis of the Concept of Physical Law', *Philosophical Review*, 1942, 51; J. Needham, *The Grand Titration*, London, Allen & Unwin, 1969, Ch. 8; J. R. Milton, 'The Origin and Development of the Concept of the "Laws of Nature" ', *Arch. europ. sociol.*, 1981, 22; Jane Ruby, 'The Origins of Scientific "Law" ', *Journal of the History of Ideas*, 1986, 42.
78 *What is a Law of Nature?*, Cambridge, Cambridge University Press, 1983.
79 Ibid., p. 77.
80 *How the Laws of Nature Lie*, Oxford, Oxford University Press, 1984.
81 'The Truth Doesn't Explain Much', *American Philosophical Quarterly*, 17, 1980, p. 161.
82 *The Scientific Image*, Oxford, Clarendon Press, 1980, p. 8.
83 See Hilary Putnam, *Mathematics, Matter and Method*, Cambridge, Cambridge University Press, 1975, Vol. 1, pp. 69f.
84 'Truth and Use', in M. Platts (ed.), *Reference, Truth and Reality*, London, Routledge, 1980, p. 19.
85 *Meaning and the Moral Sciences*, London, Routledge, 1978, pp. 34–5.
86 *The Logic of Scientific Discovery*, Engl. transl., London, Hutchinson, 1959, p. 79.
87 *The Structure of Scientific Revolutions*, p. 206.
88 *Remarks on the Foundations of Mathematics*, VI, 23.
89 *The Central Questions of Philosophy*, Harmondsworth, Penguin, 1973, end of chapter 'The Problem of Perception'.
90 *Truth and Other Enigmas*, Cambridge, Mass., Harvard University Press, 1978, p. 146.
91 *From a Logical Point of View*, 2nd edn, New York, Harper & Row, 1963, p. 44.
92 Op. cit., p. 12. Italics in the original.
93 Ibid., p. 39.
94 *Truth and Other Enigmas*, op. cit., p. 146.
95 Op. cit., p. 360.

INDEX